AF705239

Ergodic Theory

AMS/MAA | THE CARUS MATHEMATICAL MONOGRAPHS

VOL 38

Ergodic Theory

Simon Rubinstein-Salzedo

MAA PRESS

An Imprint of the AMERICAN MATHEMATICAL SOCIETY

2024 Editorial Committee
Bruce P. Palka, Editor

Jennifer Beineke David A. Brown Annalisa Crannell
Greg Friedman Jeffrey Nunemacher Edward Scheinerman
Carol Schumacher Henry Segerman Genevieve Walsh

2020 *Mathematics Subject Classification*. Primary 37-01, 37A05, 37A44, 11K16, 11K50, 11K55.

Library of Congress Cataloging-in-Publication Data

Names: Rubinstein-Salzedo, Simon, author.
Title: Ergodic theory / Simon Rubinstein-Salzedo.
Description: Providence, Rhode Island : MAA Press, an imprint of the American Mathematical Society, [2025] | Series: The Carus mathematical monographs, 0069-0813 ; volume 38 | Includes bibliographical references and index.
Identifiers: LCCN 2025003643 | ISBN 9781470479206 (paperback) | ISBN 9781470480486 (ebook)
Subjects: LCSH: Ergodic theory–Textbooks. | Dynamics–Textbooks. | Number theory–Textbooks. | AMS: Dynamical systems and ergodic theory – Instructional exposition (textbooks, tutorial papers, etc.). | Dynamical systems and ergodic theory – Ergodic theory – Measure-preserving transformations. | Number theory – Probabilistic theory: distribution modulo 1; metric theory of algorithms – Normal numbers, radix expansions, Pisot numbers, Salem numbers, good lattice points, etc.. | Number theory – Probabilistic theory: distribution modulo 1; metric theory of algorithms – Metric theory of continued fractions.
Classification: LCC QA313 .R835 2025 | DDC 515/.48–dc23/eng/20250618
LC record available at https://lccn.loc.gov/2025003643

Copying and reprinting. Individual readers of this publication, and nonprofit libraries acting for them, are permitted to make fair use of the material, such as to copy select pages for use in teaching or research. Permission is granted to quote brief passages from this publication in reviews, provided the customary acknowledgment of the source is given.

Republication, systematic copying, or multiple reproduction of any material in this publication is permitted only under license from the American Mathematical Society. Requests for permission to reuse portions of AMS publication content are handled by the Copyright Clearance Center. For more information, please visit `www.ams.org/publications/pubpermissions`.

Send requests for translation rights and licensed reprints to `reprint-permission@ams.org`.

© 2025 by the author. All rights reserved.
Printed in the United States of America.

∞ The paper used in this book is acid-free and falls within the guidelines established to ensure permanence and durability.
Visit the AMS home page at `https://www.ams.org/`

10 9 8 7 6 5 4 3 2 1   30 29 28 27 26 25

Contents

Preface

This book is the textbook for a class I have taught twice at Euler Circle to high-school (and younger) students, once in 2019 and once in 2024. Typically, students in this class would have taken, or otherwise learned the material from, the two more elementary yearlong sequences at Euler Circle. The first of these is the transition to proofs sequence based on the book [**Rub23**], covering elementary number theory, combinatorics, and analysis. The second of these is the fundamentals of higher mathematics sequence, covering basic linear and abstract algebra, real analysis, and point-set topology. So, all these topics are potentially assumed as background for the ergodic theory class, although I review some of the material that students might not have mastered fully the first time they saw it.

The first nine chapters of this book each correspond to one week of class at Euler Circle, which consists of one 2-hour lecture and one 2-hour problem session. Each week, students submit their problem sets to their TAs and go over them 1-on-1 during the problem session, so they get feedback and mentorship throughout the class, and of course I am always available to discuss mathematics with my students. There are far too many problems for students to be reasonably expected to solve, or even work on, all of them, although there were several students who did try to solve all of them and succeeded at solving most of them. They are also at a wide range of difficulty levels, from relatively basic concept-check questions to difficult theorems from the literature, and even a handful of open problems. So, the reader should not feel obligated to solve all the problems either. Being able to do roughly half the problems per chapter indicates a fairly strong understanding of the material.

Students in my class also write expository papers of roughly 5 pages on some topic of their choice related to the material but not directly covered in class, although some of them are alluded to in the problem sets. Some of the topics I suggested were the Baire Category Theorem, the Stone–Weierstraß Theorem, generalized Lüroth series, the ergodic theory of non-integer bases, natural extensions of measure-preserving systems, probability from the measure-theoretic viewpoint, L^p spaces, Haar measure, ultrafilters and their applications, spectral techniques in ergodic theory, Fourier-analytic techniques in ergodic theory, the Mandelbrot set, symbolic dynamics, and rational billiards. For this, students have to seek out resources, determine their focus and scope, and then tell their story nicely. I gained a lot from such assignments when I was an undergraduate and graduate student, both in terms of learning how to assimilate mathematical knowledge from papers and learning how to write well, especially coming up with my mathematical writing style, which my students and other readers seem to appreciate. I encourage other instructors who teach a class out of this book to include a similar assignment as part of the class.

I committed to teaching ergodic theory the first time after developing a bit of familiarity with the basic ideas of the subject, but before coming up with a good narrative for the subject and figuring out what I wanted to focus on. The first sources I read in depth were [**DK02**] and [**Sil08**], so my understanding of the subject is heavily influenced by these two books. Both times I have taught this class, it has been the advanced class in the winter quarter, while the fall and spring advanced classes were on number theory (covering reciprocity theorems, quadratic forms, and elliptic curves) and analytic number theory, respectively, so this class was part of a year that was otherwise focused on number theory. Thus it made sense to include several number-theoretic applications of ergodic theory, such as normal numbers and continued fractions, as well as Szemerédi's Theorem. So, this was one of my main motivations throughout the book and the class.

I am grateful to all my students and teaching assistants who gave me corrections and other suggestions for making this book better. Special thanks also to Alex Altair, Bryna Kra, Aaron Lin, Ho Chung Siu, Adeline Wong, and the anonymous referees, who made many useful comments and suggestions. I'm particularly indebted to Ravi Vakil, who surprised me by telling me how much he liked an earlier version of this book and encouraging me to try to publish it. Finally, I wish to thank the staff at the American Mathematical Society who made it possible to get this book to publication.

Introduction

This is a book on ergodic theory, but if you read the first three chapters, it might not feel like it. So, before you do that, let me say a bit about what ergodic theory is, and why we need to introduce a substantial amount of background from measure theory before we get started on ergodic theory.

0.1 What is Ergodic Theory?

Ergodic theory is a subfield of dynamical systems. The setup for dynamical systems is that we have a set X and a function $T : X \to X$. Because the domain and the codomain are the same set, we can iterate T, and given some $x \in X$, we can consider the sequence $x, T(x), T^2(x), T^3(x), \ldots$ There are many directions we can go from here, by considering special types of sets X and functions T.

For instance, X might be $\mathbb{C}$ and T an analytic function. The study of these sorts of dynamical systems is known as *complex dynamics*; see for instance [**CG93**] for a book on this subject. Or perhaps X is $\mathbb{Q}$ or $\mathbb{Q} \cup \{\infty\}$ and T is a rational function. The study of these dynamical systems is known as *arithmetic dynamics*; see [**Sil07**] for a book on this subject.

In ergodic theory, we specialize in a different direction. We want to consider sets X that admit a notion of size, such as length or area, or perhaps some more abstract notion of size. Given a subset $A \subseteq X$, we let $\mu(A)$ denote the size, or *measure*, of A. In general, when we consider functions on a set X with additional structure (such as a measure), we frequently want to restrict to functions that preserve this additional structure. In the case of dynamical systems on a *measure space* X, we want our function $T : X \to X$ to respect the measure, which means something like $\mu(T(A)) = \mu(A)$ for all subsets $A \subseteq X$. This is the correct notion when T is bijective, but we need a slight modification when T is not bijective; as we will see in Chapter 4, the relevant functions T satisfy $\mu(T^{-1}(A)) = \mu(A)$, where $T^{-1}(A) = \{x \in X : T(x) \in A\}$.

0.2 Motivating Examples

One of the most important measure spaces X we will be considering in this book is $\mathbb{R}/\mathbb{Z}$, or (more or less equivalently) $[0, 1)$. One of the measure-preserving transformations we will be looking at repeatedly is $T : \mathbb{R}/\mathbb{Z} \to \mathbb{R}/\mathbb{Z}$ given by $T(x) = x + \alpha \pmod 1$, where α is some fixed real number. If we picture $\mathbb{R}/\mathbb{Z}$ as a circle, then T is a rotation by α on the circle. In this case, it is clear that T preserves length: if I is any interval in $\mathbb{R}/\mathbb{Z}$, or arc on the circle, then $T^{-1}(I)$ is another arc of the same length, but rotated by $-\alpha$. We could probably manage this example without developing measure theory to the extent that we do.

But another transformation we will be looking at repeatedly is $T : \mathbb{R}/\mathbb{Z} \to \mathbb{R}/\mathbb{Z}$ given by

$$T(x) = \begin{cases} \frac{1}{x} \pmod 1 & x \neq 0, \\ 0 & x = 0. \end{cases}$$

This transformation turns out to be what we need to study the statistical properties of continued fractions. If I is an interval in $\mathbb{R}/\mathbb{Z}$, then $T^{-1}(I)$ is a union of infinitely many intervals. Furthermore, T is *not* measure-preserving with respect to the usual measure; we need to introduce a new measure for which it is measure-preserving. For this transformation and measure, it would be rather awkward to analyze it without considerable background from measure theory.

0.3 The Power of Ergodic Theory

One test of the importance of an area of mathematics is its ability to answer questions that could be asked but not easily answered without it. Ergodic theory provides powerful tools for answering statistical questions about "typical" elements of a set X. For example, given a "typical" real number x, what proportion of the digits in its decimal expansion are 4's? Once we have proven some of the key theorems in ergodic theory, we will see that the answer is $\frac{1}{10}$ for "most" real numbers x; while there are infinitely many, and even uncountably many, exceptional real numbers x that have a different proportion of 4's, or no proportion at all, they are very sparse in a measure-theoretic sense. We will similarly be able to determine the proportion of terms in the continued fraction expansion of typical real numbers that are equal to, say, 19.

In the 1970's, Furstenberg discovered another remarkable application of ergodic theory, to additive combinatorics. One of the cornerstone results in additive combinatorics is Szemerédi's Theorem, which says that sufficiently large subsets of the positive integers must contain arbitrarily long arithmetic progressions. This turns out to be equivalent to a theorem of Furstenberg that is an extension of a classic theorem of Poincaré which was the first major result in ergodic theory.

Another direction that ergodic theory takes us is to information theory and entropy. One problem we'll explore here is to analyze the information contained in a typical digit of the decimal expansion of a number, versus a typical term of the continued fraction expansion of the same number.

0.4 Structure of the Book

We begin this book by developing enough measure theory for our purposes in the rest of the book. This comprises Chapters 1–3. After that, we cover the key concepts and results in basic ergodic theory in Chapters 4–7. We begin a hierarchy of behavior for the types of transformations we're interested in in ergodic theory. The least restrictive type of transformation is the measure-preserving transformation. The next tier, which is somewhat more restrictive, is the ergodic transformation. We continue this hierarchy in Chapter 8 by introducing two more tiers, which are increasingly restrictive: weakly mixing and mixing transformations. In Chapters 9 and 10 we look at some of Furstenberg's work and its applications to additive combinatorics. The proof of Szemerédi's Theorem is beyond the scope of this book, but we instead prove a result of similar spirit that uses some of the same ideas, but that is considerably easier, namely the polynomial recurrence theorem. Finally, we discuss entropy in Chapter 11.

0.5 Prerequisites

I have tried to keep the prerequisites for this book to a minimum. Readers should certainly be comfortable with proof-based mathematics, but beyond that, there is little that is formally required. Some familiarity with real analysis, in particular comfort with arguments involving ε, is helpful, but not strictly necessary. Calculus is also helpful, but again not strictly necessary: while we will be using integrals on a regular basis, we'll develop everything we need about them, especially the measure-theoretic form of integration (the Lebesgue integral) from scratch. A few of the students I've had in classes I have taught out of this book had not seen calculus or real analysis and were still able to understand the material and solve problems.

Chapter 1

Introduction to Measure Theory

1.1 Measurable Difficulties

The point of measure theory is to be able to assign things like lengths, areas, and volumes to sets. For some sets, this is easy: for example, if we have an interval in $\mathbb{R}$, such as $[a, b]$ or (a, b) where $a \leq b$, then its length is clearly $b - a$. But for some other sets, it might not be so obvious what its length is. For instance, what is the length of $[0, 1] \cap \mathbb{Q}$, the rational numbers from 0 to 1?

Actually, this question should not be that hard either: there are countably many rational numbers between 0 and 1, as opposed to uncountably many irrationals. Since there are way more irrationals than rationals, the length of the rationals ought to be 0. More generally, any countable set ought to have length 0.

There are four basic properties that we would like a measure μ on $\mathbb{R}$ to satisfy:

(1) μ is a function from the power set of $\mathbb{R}$, i.e. the set of all subsets of $\mathbb{R}$, to $[0, \infty]$. (We allow a set to have infinite measure, because, for example, $\mathbb{R}$ itself has infinite length.)

(2) If $I \subseteq \mathbb{R}$ is an interval (open, closed, or half-open), then $\mu(I)$ is the length of I.

(3) If $X_1, X_2, \ldots$ is a finite or countable collection of subsets of $\mathbb{R}$ such that $X_i \cap X_j = \varnothing$ for all $i \neq j$, then

$$\mu\left(\bigcup_n X_n\right) = \sum_n \mu(X_n).$$

The sum can be infinite if it's a divergent series of nonnegative numbers, or if any $\mu(X_n)$ is infinity.

(4) μ is translation invariant. If $X \subseteq \mathbb{R}$ and $a \in \mathbb{R}$, then define $X + a = \{x + a : x \in X\}$. We would like $\mu(X + a) = \mu(X)$ for all X and a.

Unfortunately, we have the following issue:

Proposition 1.1. *There are no functions μ satisfying the four properties above.*

Proof. Note that the properties imply another property: monotonicity. This means that if $A \subseteq B$, then $\mu(A) \leq \mu(B)$. This is because we have $B = A \cup (B \setminus A)$, so $\mu(B) = \mu(A) + \mu(B \setminus A) \geq \mu(A)$ since $\mu(B \setminus A) \in [0, \infty]$. Notice also that if $x \in \mathbb{R}$, then $\mu(\{x\}) = 0$, because if we have any $y > x$, then $\mu((x, y)) = \mu([x, y)) = \mu((x, y)) + \mu(\{x\})$, so $\mu(\{x\}) = 0$.

We now show that properties (1)–(4) cannot be simultaneously satisfied. Suppose to the contrary that such a function μ were to exist. Define an equivalence relation $\sim$ on $\mathbb{R}$ by stating that $x \sim y$ if and only if $x - y \in \mathbb{Q}$. Each equivalence class contains some element in $[0, 1]$; actually it contains lots of them, but choose one at random or in any way, and let $E \subseteq [0, 1]$ be a set that contains exactly one representative from each equivalence class. (We need the Axiom of Choice to guarantee the existence of such a set E, and indeed without the Axiom of Choice it is not possible to prove Proposition 1.1.) Let $r_1, r_2, r_3, \ldots$ be a list of all the rational numbers in the interval $[-1, 1]$. For each i, consider the set $E_i = E + r_i$. If $i \neq j$, then $E_i \cap E_j = \varnothing$. They are all translates of E, so they all have the same measure $\mu(E)$ as E. Furthermore, $[0, 1] \subseteq \bigcup_i E_i \subseteq [-1, 2]$. Thus we have

$$1 = \mu([0, 1]) \leq \mu\left(\bigcup_{i=1}^{\infty} E_i\right) = \sum_{i=1}^{\infty} \mu(E_i) = \sum_{i=1}^{\infty} \mu(E) \leq 3.$$

But there is no number $\mu(E)$ such that $1 \le \sum_{i=1}^{\infty} \mu(E) \le 3$. Thus no such function μ exists. ■

That means that we need to drop at least one of the properties. We choose to drop property (1), that μ is defined on all the subsets of $\mathbb{R}$, and instead allow it only to be defined on *some* of the subsets, but with certain rules about which subsets are allowable.

1.2 Measure Spaces

Now that we have seen the trouble we can get into if we try to define measures too carelessly, we turn to the precise definitions.

Definition 1.2. Let X be a set. A *σ-algebra* on X is a set $\mathcal{A}$ of subsets of X such that

- $\varnothing \in \mathcal{A}$,
- If $S \in \mathcal{A}$, then $S^c \in \mathcal{A}$, where $S^c = X \setminus S$ is the complement of S,
- If $S_1, S_2, \dots$ is a finite or countable collection of elements of $\mathcal{A}$, then
$$\bigcup_n S_n \in \mathcal{A} \qquad \text{and} \qquad \bigcap_n S_n \in \mathcal{A}.$$

We call the elements of $\mathcal{A}$ the *measurable sets*, or $\mathcal{A}$-measurable sets when the σ-algebra $\mathcal{A}$ might not be clear from context.

In fact, closure under countable intersections follows from closure under countable unions (or vice versa), since
$$\bigcap_n S_n = \left(\bigcup_n S_n^c\right)^c.$$
Thus, in checking that $\mathcal{A}$ is a σ-algebra, we only have to check closure under either unions or intersections, and not both.

Definition 1.3. Given a set X and a σ-algebra on X, a *measure* on $(X, \mathcal{A})$ is a function $\mu : \mathcal{A} \to [0, \infty]$ such that the following properties hold:

- $\mu(\varnothing) = 0$,

- If $S_1, S_2, \ldots$ is a finite or countable collection of pairwise disjoint elements of $\mathcal{A}$, then

$$\mu\left(\bigcup_n S_n\right) = \sum_n \mu(S_n).$$

Definition 1.4. If X is a set, $\mathcal{A}$ is a σ-algebra on X, and μ is a measure on $(X, \mathcal{A})$, then the triple $(X, \mathcal{A}, \mu)$ is called a *measure space.*

Example 1.5. Let X be a finite or countable set. Then $\mathcal{A} = \mathcal{P}(X)$, the power set of X, is a σ-algebra on X. The *counting measure* μ on X is defined to be $\mu(S) = |S|$, the number of elements in S (which is ∞ if S is an infinite set). This is a measure, so $(X, \mathcal{A}, \mu)$ is a measure space.

Example 1.6. If X is a finite set with $|X| = n$, then we can modify the counting measure on $(X, \mathcal{P}(X))$ by setting $\mu'(A) = \frac{|A|}{n}$ for any $A \subseteq X$. This is again a measure, but now we have $\mu'(X) = 1$.

Definition 1.7. If $(X, \mathcal{A}, \mu)$ is a measure space and $\mu(X) = 1$, then μ is said to be a *probability measure* and $(X, \mathcal{A}, \mu)$ is said to be a *probability space.*

The terminology comes from the idea that we can pick a random element of X, and the probability that we land in a subset $S \subseteq X$ is $\mu(S)$. Measure theory wasn't originally developed for use in probability, but it turns out to be exactly the right setting for doing probability. This isn't a book on probability, but many ideas from probability are also relevant to ergodic theory, so it will be helpful to develop a bit of probability from time to time.

Example 1.8. Let X be any nonempty set, and let $x \in X$. Let us take $\mathcal{A} = \mathcal{P}(X)$. Define a measure δ_x on X by setting

$$\delta_x(S) = \begin{cases} 1 & x \in S, \\ 0 & x \notin S. \end{cases}$$

This is a measure, sometimes called a *point mass.*

Sets with measure 0 play a special role in measure theory and ergodic theory: we'll typically be able to ignore them. It's helpful to give them a special name.

Definition 1.9. A set of measure 0 is called a *null set*.

We'll also sometimes come across properties that hold for all points in a measure space X, except for a null set. When this happens, we say that the property holds *almost everywhere*, or for *almost all* points. In general, the phrases "almost everywhere," "almost all," and also "almost surely" mean "except for a null set." For a typical example, we might say that a function is continuous almost everywhere. A typical example of this is the Riemann function $f : \mathbb{R} \to \mathbb{R}$, defined by

$$f(x) = \begin{cases} \frac{1}{q} & x = \frac{p}{q} \text{ in lowest terms,} \\ 0 & x \in \mathbb{R} \setminus \mathbb{Q}. \end{cases}$$

One can check that f is continuous at all irrational numbers but no rational numbers. In a typical measure space like the Borel or Lebesgue space, the set of discontinuous points $\mathbb{Q}$ has measure 0, so we can say that this function is continuous almost everywhere.

Another typical example comes from probability. Suppose you flip a fair coin n times, where n is large. Then you *might* get heads far more than half the time, but that would be a highly unusual event. Indeed, the proportion of heads you get tends to $\frac{1}{2}$ almost surely. (This requires a proof, and it follows from any reasonable version of the *Law of Large Numbers*, which we will return to in Chapter 6 when we study the Birkhoff Ergodic Theorem.)

1.3 The Borel σ-algebra

A useful starting point of a σ-algebra will be the Borel σ-algebra on $\mathbb{R}$. It isn't the most useful σ-algebra we can put on $\mathbb{R}$ (that will be the *Lebesgue σ-algebra*, which we'll talk more about in Chapter 2), but it's a good first attempt at a σ-algebra on $\mathbb{R}$. Then we'll put a measure on the Borel σ-algebra, which we will later extend to the Lebesgue σ-algebra.

First, we need to start with open sets.

Definition 1.10. A subset $A \subseteq \mathbb{R}$ is said to be *open* if, for every point $x \in A$, there is some $r > 0$ such that the interval $(x - r, x + r) \subseteq A$.

There is a very simple description of the open subsets of $\mathbb{R}$:

Proposition 1.11. *A subset of $\mathbb{R}$ is open if and only if it is the disjoint union of a finite or countable number of disjoint open intervals, where open intervals are allowed to go to $-\infty$ or $+\infty$.*

Definition 1.12. The *Borel σ-algebra* on $\mathbb{R}$ is the smallest σ-algebra on $\mathbb{R}$ containing all the open sets. Any set in the Borel σ-algebra is called a *Borel set*.

Let's take a look at what sorts of sets are Borel sets. First, since σ-algebras are closed under complements, the complement of any open set is Borel.

Definition 1.13. A set is said to be *closed* if its complement is open.

Example 1.14. A closed interval $[a, b]$ is closed, because its complement $(-\infty, a) \cup (b, \infty)$ is open.

Example 1.15. Another famous closed set is the *Cantor set*. It is defined by starting with $C_0 = [0, 1]$. Then, in every iteration, we remove the open middle third of each interval. Thus $C_1 = [0, 1] \setminus \left(\frac{1}{3}, \frac{2}{3}\right) = \left[0, \frac{1}{3}\right] \cup \left[\frac{2}{3}, 1\right]$, $C_2 = \left[0, \frac{1}{9}\right] \cup \left[\frac{2}{9}, \frac{1}{3}\right] \cup \left[\frac{2}{3}, \frac{7}{9}\right] \cup \left[\frac{8}{9}, 1\right]$, and so forth. The Cantor set is defined to be $C = \bigcap_{n=0}^{\infty} C_n$. This set is closed because its complement is $(-\infty, 0) \cup (1, \infty)$, unioned with all the countably many open intervals we removed. Concretely, the Cantor set is the set of numbers in $[0, 1]$ that have a base-3 expansion containing only 0's and 2's.

All open and closed sets are Borel, but the Borel sets contain far more than just those. For example, we can get a half-open interval: $[0, 1)$ is Borel because we have $[0, 1) = \left[0, \frac{1}{2}\right] \cup (0, 1)$ for instance: the union of two (or any finite number, or even countably many) Borel sets is Borel.

Another example of a Borel set is $\mathbb{Q}$, the set of rational numbers. This is because any point $\{x\}$ is closed, and $\mathbb{Q}$ contains countably many points. So we have

$$\mathbb{Q} = \bigcup_{q \in \mathbb{Q}} \{q\},$$

a countable union of Borel sets. Similarly, the set of irrational numbers is Borel, because it's the complement of $\mathbb{Q}$, which we already know to be Borel.

Notice that $\mathbb{Q}$ is a union of countably many closed sets, but it is not itself closed (because its complement $\mathbb{R}\setminus\mathbb{Q}$ is not open). We have a name for such a set.

Definition 1.16.

- A set is said to be an F_σ set if it is a countable union of closed sets.
- A set is said to be a G_δ set if it is a countable intersection of open sets.

Remark 1.17. The mnemonic here is that σ stands for sum, and a union is rather like a sum. On the other hand, δ stands for difference, and an intersection is a bit like a difference.

All F_σ and G_δ sets are Borel sets. Clearly, all closed sets are F_σ sets and all open sets are G_δ sets; for instance, if C is closed, then let $C_n = C$ for all n, so that $C = \bigcup_n C_n$, so this shows that closed sets are F_σ's. A similar argument shows that open sets are G_δ's. Furthermore, open sets are also F_σ's, which is a bit less obvious. To see this, recall that an open set is an at-most-countable disjoint union of open intervals. If we can show that one open interval is an F_σ, then an arbitrary open set is a countable union of countable unions of closed sets, and hence a countable union of closed sets.

So, let's show that an open interval is an F_σ set. We'll work with the open interval $(-1, 1)$. We just have to write it as a countable union of closed sets, so let's do it:

$$(-1,1) = \bigcup_{n=1}^{\infty}\left[-1+\frac{1}{n}, 1-\frac{1}{n}\right].$$

A similar argument can be used to show that closed sets are G_δ's.

An example of a set that is *not* an F_σ is $\mathbb{R}\setminus\mathbb{Q}$, the set of irrational numbers. However, this is not obvious, and it relies on a nontrivial theorem called the Baire Category Theorem. Its proof is by no means beyond our level of ability to do, but as it is somewhat off-topic, let us skip the proof.

Similarly, there are sets that are not G_δ's. A typical example is $\mathbb{Q}$. This follows easily once we know that $\mathbb{R}\setminus\mathbb{Q}$ is not an F_σ, and vice versa.

Knowing examples of Borel sets that are not F_σ's and G_δ's, we can construct a Borel set that is neither an F_σ nor a G_δ. A typical example is $S = (\mathbb{Q} \cap (-\infty, 0]) \cup ((\mathbb{R} \setminus \mathbb{Q}) \cap (0, \infty))$.

There are also sets that are not Borel. They require a bit more work to define, but the starting point is the set E we described in the proof of Proposition 1.1.

1.4 Outer Measures

We would like to put a measure on the Borel sets of $\mathbb{R}$ such that the measure of any interval is its length. In order to do that, we have to start by constructing a measure-like object on all subsets of $\mathbb{R}$. This won't have all the properties of a measure, but when we restrict to the Borel subsets, then it will.

Definition 1.18. Let X be a set. An *outer measure* on X is a function $\mu^* : \mathcal{P}(X) \to [0, \infty]$ satisfying the following properties:

- $\mu^*(\varnothing) = 0$.
- If $A \subseteq B \subseteq X$, then $\mu^*(A) \le \mu^*(B)$. This property is called *monotonicity*.
- If $A_1, A_2, \ldots \subseteq X$, then

$$\mu^*\left(\bigcup_{n=1}^{\infty} A_n\right) \le \sum_{n=1}^{\infty} \mu^*(A_n).$$

  This property is called *countable subadditivity*.

Example 1.19. Let X be any set. Define μ^* by

$$\mu^*(A) = \begin{cases} 0 & A = \varnothing, \\ 1 & A \ne \varnothing. \end{cases}$$

Then μ^* is an outer measure.

Example 1.20. Let X be any set. Define μ^* by

$$\mu^*(A) = \begin{cases} 0 & A \text{ is countable}, \\ 1 & A \text{ is uncountable}. \end{cases}$$

Then μ^* is an outer measure.

The most relevant outer measure for our purposes is the *Lebesgue outer measure.* It is defined as follows.

Definition 1.21. The *Lebesgue outer measure* λ^* on $\mathbb{R}$ is defined as follows. For any $A \subseteq \mathbb{R}$, let $\mathscr{C}(A)$ denote the set of all finite or countable sequences (x_i, y_i) of open intervals in $\mathbb{R}$ such that $A \subseteq \bigcup_i (x_i, y_i)$. Then define

$$\lambda^*(A) = \inf\left\{\sum_i (y_i - x_i) : \{(x_i, y_i)\} \in \mathscr{C}(A)\right\}.$$

Intuitively, what this means is that we cover the set A by a finite or countable number of open intervals and then take the sum of the lengths of these intervals. Then we take the infimum over all possible values obtained by various coverings of A.

Remark 1.22. The *infimum* of a set S of real numbers is defined to be the *greatest lower bound* of S, i.e. the largest real number r such that $r \leq s$ for all $s \in S$. It's similar to a minimum, except that if S is infinite, then it might not have a minimum. For instance, $\{x : x > 0\}$ has no minimum, but it has an infimum, namely 0.

It isn't immediately obvious that the Lebesgue outer measure λ^* is in fact an outer measure. However, this does turn out to be true.

Proposition 1.23. *The Lebesgue outer measure λ^* is an outer measure on $\mathbb{R}$. Furthermore, if $I \subseteq \mathbb{R}$ is an interval, then $\lambda^*(I)$ is the length of I.*

Proof. We first check that λ^* is an outer measure. The first thing we must show is that $\lambda^*(\varnothing) = 0$. This is because it is covered by a union of no intervals, so the sum of their lengths is 0. Since λ^* only takes on nonnegative values, we have $\lambda^*(\varnothing) = 0$.

Next, we must show monotonicity. Suppose $A \subseteq B \subseteq \mathbb{R}$. Then $\mathscr{C}(B) \subseteq \mathscr{C}(A)$ since any collection of open intervals that covers B automatically also covers A. If we take an infimum over a larger set, the resulting value can only decrease. Thus $\lambda^*(A) \leq \lambda^*(B)$.

Next, we must show countable subadditivity. Suppose $A_1, A_2, \ldots \subseteq \mathbb{R}$, and let $\varepsilon > 0$. For each n, pick an open cover $\{(x_i^n, y_i^n)\}$ in $\mathscr{C}(A_n)$ such that $\sum_i (y_i^n - x_i^n) \le \lambda^*(A_n) + \frac{\varepsilon}{2^n}$. Then $T = \{(x_i^n, y_i^n) : i \ge 1, n \ge 1\} \in \mathscr{C}(\bigcup_n A_n)$, and we have

$$\begin{aligned}\sum_{(x,y)\in T} (y - x) &\le \sum_{i,n} (y_i^n - x_i^n) \\ &= \sum_n \left(\sum_i (y_i^n - x_i^n) \right) \\ &\le \sum_n \left(\lambda^*(A_n) + \frac{\varepsilon}{2^n} \right) \\ &= \sum_n \lambda^*(A_n) + \varepsilon.\end{aligned}$$

Since the sum on the left is one of the numbers in the infimum, it follows that $\lambda^*(A) \le \sum_n \lambda^*(A_n) + \varepsilon$. Since ε is arbitrary, it follows that $\lambda^*(A) \le \sum_n \lambda^*(A_n)$.

Finally, suppose that I is an interval, say with endpoints a and b. (It can be open, closed, or half-open, and the argument still goes through.) For any $\varepsilon > 0$, I is contained in $(a - \varepsilon, b + \varepsilon)$, which has length $b - a + 2\varepsilon$. Since this is one of the items in the infimum in the definition of $\lambda^*(I)$, we have $\lambda^*(I) \le b - a + 2\varepsilon$ for all $\varepsilon > 0$, which implies that $\lambda^*(I) \le b - a$. Now, cover I by any finite or countable family of open intervals (x_i, y_i) such that the sum of the lengths of these intervals is finite. Pick $\varepsilon > 0$, and find an n such that the sum of the lengths of the subintervals of I not covered by the first n intervals $(x_1, y_1), \dots, (x_n, y_n)$ is less than ε, i.e. the sum of the lengths of the subintervals in

$$I \setminus \bigcup_{i=1}^n (x_i, y_i)$$

is less than ε. Then $\sum_{i=1}^n (y_i - x_i)$ must be at least $(b - a) - \varepsilon$. Since this holds for any $\varepsilon > 0$, we must have $\sum_{i=1}^\infty (y_i - x_i) \ge b - a$. Thus $\lambda^*(I) \ge b - a$, as desired. ■

In much the same way, we can define an outer measure, which we'll also call λ^*, on $\mathbb{R}^n$ for any n; the dimension should always be clear from context. For a subset $A \subseteq \mathbb{R}^n$, cover A by finitely many or countably many open rectangles with sides parallel to the coordinate axes, then take the sum of their areas/volumes. We define $\lambda^*(A)$ to be the infimum over all such sums of areas/volumes.

1.5 Measures from Outer Measures

Let X be a set and μ^* an outer measure on X. There is a way of defining an associated σ-algebra $\mathcal{A}$ on X and a measure μ on $(X, \mathcal{A})$ out of μ^*.

Definition 1.24. Let X be a set and μ^* an outer measure on X. We say that a subset $B \subseteq X$ is *μ^*-measurable* if, for every $A \subseteq X$, we have

$$\mu^*(A) = \mu^*(A \cap B) + \mu^*(A \cap B^c).$$

If we think of μ^* as being a proxy for length/area/volume, then this makes a lot of sense: $\mu^*(A)$ measures how much "stuff" is in A, and if we split A into two disjoint subsets $A \cap B$ and $A \cap B^c$, then we ought to have $\mu^*(A) = \mu^*(A \cap B) + \mu^*(A \cap B^c)$. However, there isn't anything in the definition of outer measure that forces this to be true, and indeed there are sometimes subsets $B \subseteq X$ for which it fails on some choice of A.

What the definition of outer measure *does* imply is that $\mu^*(A) \leq \mu^*(A \cap B) + \mu^*(A \cap B^c)$; this just follows from the subadditivity required of all outer measures. So it is only the other direction, $\mu^*(A) \geq \mu^*(A \cap B) + \mu^*(A \cap B^c)$ that we really need to check.

Proposition 1.25. *Suppose that $\mu^*(B) = 0$ or $\mu^*(B^c) = 0$. Then B is μ^*-measurable.*

Proof. Suppose that $\mu^*(B) = 0$; the case $\mu^*(B^c) = 0$ is very similar. Let $A \subseteq X$ be arbitrary. Since $\mu^*(B) = 0$, by monotonicity we also have $\mu^*(A \cap B) = 0$. Furthermore, by monotonicity, $\mu^*(A) \geq \mu^*(A \cap B^c)$. Thus we have $\mu^*(A) \geq \mu^*(A \cap B) + \mu^*(A \cap B^c)$, as desired. ∎

Theorem 1.26. *Let X be a set and μ^* an outer measure on X. Let $\mathcal{A}$ denote the set of μ^*-measurable subsets of X. Then $\mathcal{A}$ is a σ-algebra, and the restriction of μ^* to $\mathcal{A}$ is a measure on $(X, \mathcal{A})$.*

Proof. We first show that $\mathcal{A}$ is a σ-algebra. The first thing to check is that $\varnothing \in \mathcal{A}$. This is true because $\mu^*(\varnothing) = 0$, so the result follows from

Proposition 1.25. Next, suppose that $B \in \mathcal{A}$. We must show that $B^c \in \mathcal{A}$. Let $A \subseteq X$ be any subset. Since B is measurable, we have

$$\mu^*(A) = \mu^*(A \cap B) + \mu^*(A \cap B^c).$$

This is exactly the same condition needed for the μ^*-measurability of B^c, so we conclude that $B^c \in \mathcal{A}$ as well.

The last thing that we need to check to conclude that $\mathcal{A}$ is a σ-algebra, is that $\mathcal{A}$ is closed under countable unions and intersections. The result for intersections follows from unions and complements, so we just focus on unions. We first show that $\mathcal{A}$ is closed under unions of two (and hence finitely many) sets. Suppose $B_1, B_2 \in \mathcal{A}$. Since B_1 is μ^*-measurable, we have

$$\begin{aligned}\mu^*(A \cap (B_1 \cup B_2)) &= \mu^*(A \cap (B_1 \cup B_2) \cap B_1) \\ &\quad + \mu^*(A \cap (B_1 \cup B_2) \cap B_1^c) \\ &= \mu^*(A \cap B_1) + \mu^*(A \cap B_1^c \cap B_2).\end{aligned}$$

Then, first by the measurability of B_2 and then by the measurability of B_1, we have

$$\begin{aligned}&\mu^*(A \cap (B_1 \cup B_2)) \\ &+\mu^*(A \cap (B_1 \cup B_2)^c) = \mu^*(A \cap B_1) + \mu^*(A \cap B_1^c \cap B_2) \\ &\qquad\qquad + \mu^*(A \cap B_1^c \cap B_2^c) \\ &\qquad = \mu^*(A \cap B_1) + \mu^*(A \cap B_1^c) \\ &\qquad = \mu^*(A).\end{aligned}$$

Thus $B_1 \cup B_2 \in \mathcal{A}$.

Next, we must show that $\mathcal{A}$ is closed under countable unions. Let $B_1, B_2, \ldots$ be a countable collection of *pairwise disjoint* elements of $\mathcal{A}$. We will show by induction that

$$\mu^*(A) = \sum_{i=1}^{n} \mu^*(A \cap B_i) + \mu^*\left(A \cap \left(\bigcap_{i=1}^{n} B_i^c\right)\right). \tag{1.1}$$

When $n = 1$, this is simply the measurability of B_1. Now, suppose that (1.1) is true for n, and we'll prove it for $n + 1$. By the measurability of B_{n+1} and the fact that the B_i's are pairwise disjoint, we have

$$\begin{aligned}\mu^*\left(A\cap\left(\bigcap_{i=1}^{n}B_i^c\right)\right) &= \mu^*\left(A\cap\left(\bigcap_{i=1}^{n}B_i^c\right)\cap B_{n+1}\right)\\ &\quad+\mu^*\left(A\cap\left(\bigcap_{i=1}^{n}B_i^c\right)\cap B_{n+1}^c\right)\\ &=\mu^*\left(A\cap B_{n+1}\right)+\mu^*\left(A\cap\left(\bigcap_{i=1}^{n+1}B_i^c\right)\right),\end{aligned}$$

which completes the induction. Now, we have

$$\begin{aligned}\mu^*(A) &= \sum_{i=1}^{n}\mu^*(A\cap B_i)+\mu^*\left(A\cap\left(\bigcap_{i=1}^{n}B_i^c\right)\right)\\ &\geq \sum_{i=1}^{n}\mu^*(A\cap B_i)+\mu^*\left(A\cap\left(\bigcap_{i=1}^{\infty}B_i^c\right)\right)\\ &= \sum_{i=1}^{n}\mu^*(A\cap B_i)+\mu^*\left(A\cap\left(\bigcup_{i=1}^{\infty}B_i\right)^c\right)\end{aligned}$$

for any n. Thus it holds when the sum goes up to ∞ as well:

$$\mu^*(A)\geq\sum_{i=1}^{\infty}\mu^*(A\cap B_i)+\mu^*\left(A\cap\left(\bigcup_{i=1}^{\infty}B_i\right)^c\right).$$

By countable subadditivity, we have

$$\begin{aligned}\mu^*(A) &\geq \sum_{i=1}^{\infty}\mu^*(A\cap B_i)+\mu^*\left(A\cap\left(\bigcup_{i=1}^{\infty}B_i\right)^c\right)\\ &\geq \mu^*\left(A\cap\left(\bigcup_{i=1}^{\infty}B_i\right)\right)+\mu^*\left(A\cap\left(\bigcup_{i=1}^{\infty}B_i\right)^c\right)\\ &\geq \mu^*(A).\end{aligned}$$

Hence we must have equality.

Now, to ensure that $\bigcup_{i=1}^{\infty}B_i$ is μ^*-measurable, we need to be able to deal with arbitrary measurable B_i, rather than disjoint ones. However, any countable union of μ^*-measurable sets can just as easily be written as a countable *disjoint* union of μ^*-measurable sets: let

$$C_i = B_1^c \cap B_2^c \cap \cdots \cap B_{i-1}^c \cap B_i.$$

Then the C_i's are disjoint and measurable, and their union is the same as that of the B_i's. Thus we may conclude that $\mathcal{A}$ is closed under countable unions.

Finally, we must show that μ^* restricted to $(X, \mathcal{A})$ is a measure. This just means checking countable additivity on disjoint sets. Let $B_1, B_2, \ldots$ be disjoint sets in $\mathcal{A}$, and let $A = \bigcup_n B_n$. Then we have

$$\mu^*\left(\bigcup_n B_n\right) \geq \sum_n \mu^*(B_n) + 0,$$

and the reverse inequality is automatic. This completes the proof. ■

Proposition 1.27. *Let λ^* be the Lebesgue outer measure on $\mathbb{R}$. Then every Borel subset is λ^*-measurable.*

Proof. It suffices to check that all intervals of the form $(-\infty, b]$ are λ^*-measurable, since the Borel σ-algebra is the smallest σ-algebra containing all such sets (see Problem 7). Let b be an arbitrary real number, and let $B = (-\infty, b]$. We need to check that for any $A \subseteq \mathbb{R}$, we have

$$\lambda^*(A) \geq \lambda^*(A \cap B) + \lambda^*(A \cap B^c).$$

In fact, we'll restrict to the case where $\lambda^*(A) < \infty$, since otherwise both sides are ∞. Let A be such a set, and let $\varepsilon > 0$ be arbitrary. Let $\{(a_n, b_n)\}$ be a sequence of open intervals covering A, whose total length is less than $\lambda^*(A) + \varepsilon$. For each n, the sets $(a_n, b_n) \cap B$ and $(a_n, b_n) \cap B^c$ are disjoint intervals whose union is (a_n, b_n). Thus we have

$$b_n - a_n = \lambda^*((a_n, b_n)) = \lambda^*((a_n, b_n) \cap B) + \lambda^*((a_n, b_n) \cap B^c).$$

Since the sequence $\{(a_n, b_n) \cap B\}$ covers $A \cap B$ and $\{(a_n, b_n) \cap B^c\}$ covers $A \cap B^c$, countable subadditivity of λ^* implies that

$$\begin{aligned}\lambda^*(A \cap B) + \lambda^*(A \cap B^c) &\leq \sum_n \lambda^*((a_n, b_n) \cap B) \\ &\quad + \sum_n \lambda^*((a_n, b_n) \cap B^c) \\ &= \sum_n (b_n - a_n) \\ &< \lambda^*(A) + \varepsilon.\end{aligned}$$

Since ε is arbitrary, it follows that $\lambda^*(A \cap B) + \lambda^*(A \cap B^c) \leq \lambda^*(A)$, as desired. This completes the proof. ■

1.6 Problems

(1) Let $(X, \mathcal{A}, \mu)$ be a measure space with $\mu(X) < \infty$. Prove that if $A, B \in \mathcal{A}$, then

$$\mu(A \cup B) = \mu(A) + \mu(B) - \mu(A \cap B).$$

(2) Let $(X, \mathcal{A}, \mu)$ be a measure space. Prove that μ is a *monotonic* function, i.e. if $A, B \in \mathcal{A}$ with $A \subseteq B$, then $\mu(A) \leq \mu(B)$.

(3) (a) Let $X = \{a, b, c\}$ be a set containing three elements. List all possible σ-algebras on X.

(b) What are the possibilities for the number of measurable sets in a σ-algebra on a finite set? (Your answer should be something like "the number of measurable sets must be prime, and it can be any prime number," although this is not the correct answer.) Prove that your answer is correct.

(4) Show, by giving an example, that the union of two σ-algebras on a set X need not be a σ-algebra.

(5) Let $(X, \mathcal{A}, \mu)$ be a measure space.

(a) Suppose $A_1, A_2, \ldots \in \mathcal{A}$, and that $A_i \subseteq A_j$ whenever $i \leq j$. Prove that

$$\mu\left(\bigcup_{k=1}^{\infty} A_k\right) = \lim_{k \to \infty} \mu(A_k).$$

(b) Suppose that $B_1, B_2, \ldots \in \mathcal{A}$, and that $B_i \supseteq B_j$ whenever $i \leq j$, and that $\mu(B_n) < \infty$ for some n. Prove that

$$\mu\left(\bigcap_{k=1}^{\infty} B_k\right) = \lim_{k\to\infty} \mu(B_k).$$

(c) Find a counterexample to part (b) when $\mu(B_n) = \infty$ for all n.

(6) Let $(X, \mathcal{A}, \mu)$ be a measure space, and $A_1, A_2, \ldots$ be measurable subsets of X such that, for any i, j with $i \neq j$, $\mu(A_i \cap A_j) = 0$. Prove that

$$\mu\left(\bigcup_{k=1}^{\infty} A_k\right) = \sum_{k=1}^{\infty} \mu(A_k).$$

(7) Prove that if $\mathcal{A}$ is any σ-algebra on $\mathbb{R}$ containing all sets of the form $(-\infty, b]$ for all $b \in \mathbb{R}$, then $\mathcal{A}$ contains all Borel sets. In this sense, the Borel σ-algebra on $\mathbb{R}$ is the smallest σ-algebra containing all the sets $(-\infty, b]$.

(8) Show, directly from the definition of the Lebesgue outer measure λ^*, that if $A \subseteq \mathbb{R}$ is countable, then $\lambda^*(A) = 0$.

(9) Let X be a set and $\mathcal{A}$ a σ-algebra on X. Let $x, y \in X$. Under what conditions on x and y are the points masses δ_x and δ_y the same measure?

(10) Let $\mathcal{A}$ be the smallest σ-algebra on $\mathbb{R}$ containing all one-point sets. Give a concrete description of the $\mathcal{A}$-measurable subsets.

(11) Does there exist a σ-algebra with a countably infinite number of members (subsets)?

(12) (a) Construct a Borel set $A \subseteq [0, 1]$ such that for each subinterval I of $[0, 1]$ with positive measure, we have $0 < \lambda^*(A \cap I) < \lambda^*(I)$.

(b) Construct a Borel set $A \subseteq \mathbb{R}$ with $\lambda^*(A) < \infty$ such that for each finite subinterval I of $\mathbb{R}$ with positive measure, we have $0 < \lambda^*(A \cap I) < \lambda^*(I)$.

(13) Let $(X, \mathcal{A}, \mu)$ be a measure space, and let $A_1, A_2, \ldots \in \mathcal{A}$ be a collection of measurable sets such that $\sum_{n=1}^{\infty} \mu(A_n) < \infty$. Let B be the set of points $x \in X$ such that $x \in A_n$ for infinitely many values of n.

Prove that $\mu(B) = 0$. (This result is known as the *first Borel–Cantelli Lemma.*)

(14) Let $(X, \mathcal{A}, \mu)$ be a probability space. We say that a collection of subsets $\{A_\alpha\}_{\alpha \in I} \subseteq \mathcal{A}$ is *independent* if, for any finite subset J of I, we have

$$\mu\left(\bigcap_{j \in J} A_j\right) = \prod_{j \in J} \mu(A_j).$$

Suppose that $A_1, A_2, \ldots$ are independent and $\sum_{n=1}^{\infty} \mu(A_n) = \infty$, and let B be the set of points $x \in X$ such that $x \in A_n$ for infinitely many values of n. Prove that $\mu(B) = 1$. (This is the *second Borel–Cantelli Lemma.*)

(15) Prove that the result of Problem 14 need not hold if the A_i's are not independent.

(16) Use the result of Problem 14 to show that if a monkey types a random character on a typewriter every second for all eternity, then with probability 1 it eventually produces the complete works of Shakespeare.[1]

[1] Do not force monkeys to type random keys on typewriters for long periods of time! Abusing animals plays no role in ergodic theory.

Chapter 2

The Lebesgue Integral

2.1 Lebesgue Sets

Recall that in Chapter 1, we defined the Lebesgue outer measure on $\mathbb{R}$, as follows: for any set $B \subseteq \mathbb{R}$, we cover B be finitely many or countably many open intervals, then sum their lengths. The Lebesgue outer measure of B is the infimum of this sum of lengths. Furthermore, a set $B \subseteq \mathbb{R}$ is *measurable* if for any set $A \subseteq \mathbb{R}$, we have

$$\lambda^*(A) = \lambda^*(A \cap B) + \lambda^*(A \cap B^c).$$

We also showed last time that the measurable subsets form a σ-algebra, which we'll call the *Lebesgue σ-algebra* $\mathcal{L}$, and that the restriction of λ^* to $\mathcal{L}$ is a measure, which we call λ. We also saw that all Borel sets are measurable, so $\mathcal{L}$ contains the Borel sets.

Before we do a more careful analysis of $\mathcal{L}$, let's pause to see an example of how λ^* works.

Example 2.1. Let us compute $\lambda^*(\mathbb{Q})$. It seems like $\lambda^*(\mathbb{Q})$ should be 0, since there are far more irrationals than rationals, and this is correct, but let's prove it to make sure, and also so that we can see the typical technique of showing a set has measure 0. Since $\mathbb{Q}$ is countable, we can order the rationals $q_1, q_2, \ldots$, where each rational appears exactly once on the list. Now, pick an $\varepsilon > 0$. Cover each q_i by an open interval of length $\frac{\varepsilon}{2^i}$, which we'll call J_i. There are some overlaps, so merge the overlapping intervals into larger intervals. The union of the J_i's is thus a countable collection of disjoint open intervals, and the sum of their lengths is at most the sum of the lengths of the J_i's, i.e. $\sum_{i=1}^{\infty} \frac{\varepsilon}{2^i} = \varepsilon$. It follows that $\lambda^*(\mathbb{Q}) \leq \varepsilon$. Since $\varepsilon > 0$ was arbitrary, it follows that $\lambda^*(\mathbb{Q}) = 0$.

A very similar argument shows that any countable set has measure 0.

This proof may seem surprising. Since $\mathbb{Q}$ is dense in $\mathbb{R}$ in the sense that for any $\alpha \in \mathbb{R}$ and any $\varepsilon > 0$, there is some $q \in \mathbb{Q}$ such that $|\alpha - q| < \varepsilon$, and we cover all the rationals by small open intervals, don't these intervals cover all the real numbers as well? So, why doesn't $\mathbb{R}$ also have measure 0?

The point is that these intervals *do not* cover all of $\mathbb{R}$. In fact, they miss most points. In Problem 8, you will find some points missed by such a covering.

Some uncountable sets also have measure 0. A typical example is the Cantor set; you will prove this in Problem 3.

In Chapter 1, we proved that if μ^* is any outer measure on a set X, and $B \subseteq X$ is a subset with $\mu^*(B) = 0$, then B is μ^*-measurable. However, not every measure-0 subset of $\mathbb{R}$ is Borel (see Problem 13). It follows that the Lebesgue σ-algebra $\mathcal{L}$, consisting of all the sets that are measurable with respect to the Lebesgue outer measure, is bigger than the Borel σ-algebra: all Borel sets are Lebesgue, but there are Lebesgue sets that are not Borel. It turns out that the Lebesgue σ-algebra is the *completion* of the Borel σ-algebra. That is, $\mathcal{L}$ is the smallest σ-algebra containing all the Borel sets and all the measure-0 sets.

Proposition 2.2. *The Lebesgue σ-algebra on $\mathbb{R}$ (or an interval thereof) is the smallest σ-algebra containing the Borel sets, together with all subsets of those Borel sets that have Lebesgue measure 0.*

This will follow from the following lemma:

Lemma 2.3. *A set $A \subseteq \mathbb{R}$ (or an interval thereof) is Lebesgue-measurable if and only if for any $\varepsilon > 0$, there is an open set (i.e. a countable union of open intervals) $U \supseteq A$, depending on ε, such that $\lambda^*(U \setminus A) < \varepsilon$.*

Proof. First, suppose that A is Lebesgue-measurable, and we'll prove that for any $\varepsilon > 0$, there is an open set $U \supseteq A$ such that $\lambda^*(U \setminus A) < \varepsilon$. So, pick some $\varepsilon > 0$. Suppose first that $\lambda(A) < \infty$. By the definition of the Lebesgue outer measure (this has nothing to do with measurability yet), there is an open set $U \supseteq A$ such that $\lambda^*(U) < \lambda^*(A) + \varepsilon$. But by the definition of measurability, we have

$$\lambda^*(U) = \lambda^*(A \cap U) + \lambda^*(A^c \cap U) = \lambda^*(A) + \lambda^*(U \setminus A)$$

since $A \subseteq U$. Thus

$$\lambda^*(U \setminus A) = \lambda^*(U) - \lambda^*(A) < \varepsilon,$$

as desired.

Now, suppose that $\lambda^*(A) = \infty$. Let $A_k = \{x \in A : |x| \le k\}$. Then each A_k is measurable, and the A_k's are increasing, with $A = \bigcup_{k=1}^{\infty} A_k$. For each k, find $U_k \supseteq A_k$ such that $\lambda^*(U_k \setminus A_k) < \frac{\varepsilon}{2^k}$, and let $U = \bigcup_{k=1}^{\infty} U_k$. Then $U \supseteq A$, and we have $U \setminus A \subseteq \bigcup_{k=1}^{\infty}(U_k \setminus A_k)$. Thus we have

$$\lambda^*(U \setminus A) \le \lambda^*\left(\bigcup_{k=1}^{\infty} U_k \setminus A_k\right) \le \sum_{k=1}^{\infty} \lambda^*(U_k \setminus A_k) < \sum_{k=1}^{\infty} \frac{\varepsilon}{2^k} = \varepsilon,$$

as desired. That completes the proof in that direction.

Now, suppose that for any $\varepsilon > 0$, there is an open set $U \supseteq A$ such that $\lambda^*(U \setminus A) < \varepsilon$; pick an $\varepsilon > 0$ and choose a suitable U. Now let $B \subseteq \mathbb{R}$ be any set. We have

$$\begin{aligned}\lambda^*(A \cap B) + \lambda^*(A^c \cap B) &\le \lambda^*(A \cap B) + \lambda^*(U^c \cap B) \\ &\quad + \lambda^*((U \setminus A) \cap B) \\ &\le \lambda^*(U \cap B) + \lambda^*(U^c \cap B) + \lambda^*(U \setminus A) \\ &< \lambda^*(B) + \varepsilon.\end{aligned}$$

Since this holds for all $\varepsilon > 0$, we have $\lambda^*(A \cap B) + \lambda^*(A^c \cap B) \le \lambda^*(B)$. Since the other direction is automatic, we have $\lambda^*(A \cap B) + \lambda^*(A^c \cap B) = \lambda^*(B)$ for all $B \subseteq \mathbb{R}$, so A is Lebesgue measurable. ∎

We now use Lemma 2.3 to prove Proposition 2.2.

Proof of Proposition 2.2. We already know that all Borel sets and sets with Lebesgue (outer) measure 0 are Lebesgue-measurable, so we only have to prove the other direction. Suppose that A is Lebesgue-measurable. For each positive integer n, use Lemma 2.3 to find an open set $U_n \supseteq A$ such that $\lambda^*(U_n \setminus A) < \frac{1}{n}$. Let $U = \bigcap_{n=1}^{\infty} U_n$. Then $U \supseteq A$, but

$$\lambda^*(U \setminus A) \le \lambda^*(U_n \setminus A) < \frac{1}{n}$$

for all integers n, so $\lambda^*(U \setminus A) = 0$. Let $N = U \setminus A$; we have just shown that N is a null set, and clearly U is a Borel set, being the intersection of open sets. (In fact, U is even a G_δ-set.) Now, $A = U \setminus N$, so it suffices to show that N is contained in a Borel set of measure 0. However, by a very similar argument to before, there is a G_δ-set V such that $V \supseteq N$ and $\lambda^*(V \setminus N) = 0$, which implies that $\lambda^*(V) = 0$. This completes the proof. ■

2.2 The Riemann Integral

We now turn to the *raison d'être* of measure theory: integration. In this section, we'll look at the first formal version of the integral: the Riemann integral. If we have a function f defined on $\mathbb{R}$ or some interval thereof, then the integral $\int_a^b f(x)\,dx$ measures the area bounded by the lines $y = 0$, $x = a$, and $x = b$, and the curve $y = f(x)$. If $f(x)$ is ever negative, then that portion is counted as having negative area. But for what we'll do for now, we'll assume that $f(x) \geq 0$ for all x. See Figure 2.1 for a picture.

Definition 2.4. Let $[a, b]$ be an interval in $\mathbb{R}$, and let $f : [a, b] \to \mathbb{R}_{\geq 0}$. Consider any partition P of $[a, b]$, i.e. a finite collection of points $t_0, t_1, \dots, t_n$ with $a = t_0 < t_1 < t_2 < \cdots < t_n$, and define

$$\mathcal{R}_P(f)_a^b = \sum_{i=1}^{n} f(t_i)(t_i - t_{i-1}).$$

We say that f is *Riemann-integrable* on $[a, b]$ if

$$\lim_{\|P\| \to 0} \mathcal{R}_P(f)_a^b$$

exists, where $\|P\| = \max(t_i - t_{i-1})$. If f is Riemann-integrable on $[a, b]$, we define its *Riemann integral* to be

$$\int_a^b f(x)\,dx = \lim_{\|P\| \to 0} \mathcal{R}_P(f)_a^b.$$

That's a bit of a mouthful, but the intuitive idea is pretty straightforward. Consider the region bounded by $y = 0$, $y = f(x)$, $x = a$, and $x = b$. We approximate it by a bunch of rectangles whose widths are

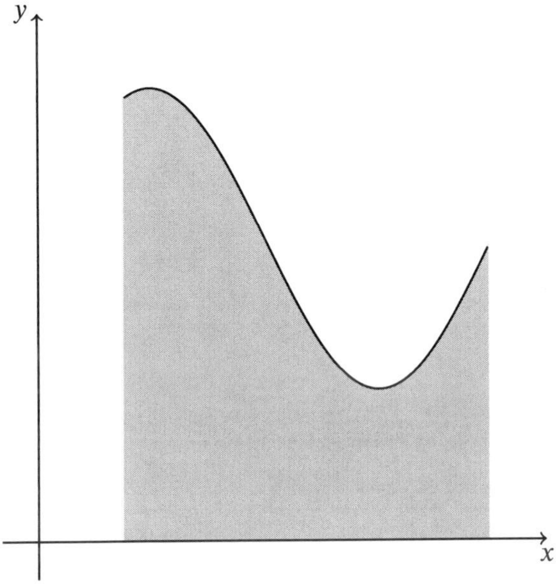

Figure 2.1. The integral is the area under the curve.

very small. If the sum of the areas of these rectangles doesn't depend much on the precise way we approximate it by rectangles, then f is integrable. (See Figure 2.2.)

Any continuous function is Riemann-integrable. We won't prove this here because we're not here to focus on that sort of thing, although it isn't particularly hard to prove this. The idea, though, is that if you've chopped up $[a, b]$ into a sufficiently fine partition, then the values of $f(x)$ on each rectangle are pretty close to each other, so if you chop it up even further still, then it won't change the value of the Riemann sum $\mathcal{R}_P(f)_a^b$ by much.

However, things can be quite different for discontinuous functions. Sometimes, discontinuous functions are still Riemann-integrable. For example, consider the function

$$f(x) = \begin{cases} 0 & x < 0, \\ 1 & x \geq 0. \end{cases}$$

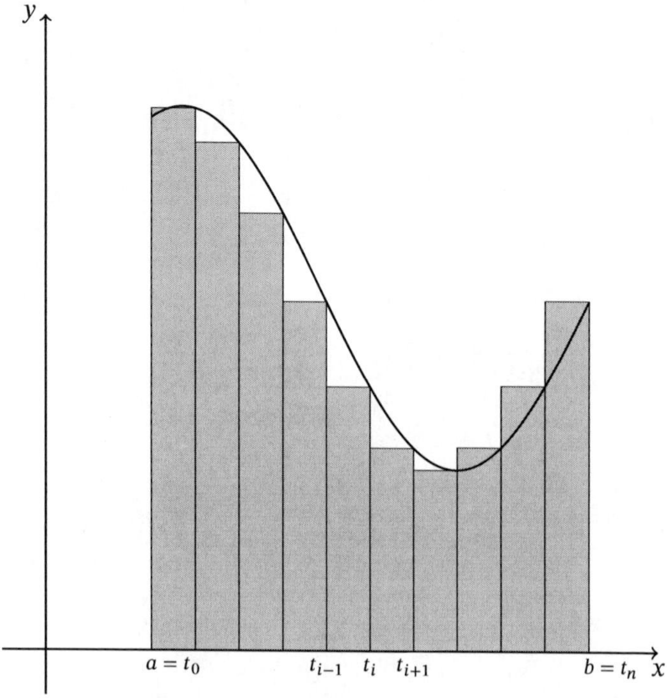

Figure 2.2. The Riemann integral

Then f is still Riemann-integrable on, say, $[-1, 1]$, because only one rectangle is affected by the discontinuity. In fact, $\int_{-1}^{1} f(x)\,dx = 1$. In fact, we have a much more powerful theorem telling us exactly when a function is Riemann-integrable:

Theorem 2.5 (Riemann–Lebesgue). *A function $f : [a, b] \to \mathbb{R}_{\geq 0}$ is Riemann-integrable iff f is bounded on $[a, b]$ (i.e. $f(x) \leq M$ for some M and for all $x \in [a, b]$) and continuous except on a null set.*

See [**Pug15**, Theorem 23, pp. 175–179] for a proof.

This concept of things happening "except on a null set" is a major theme in measure theory and ergodic theory, so let's give it a name.

Definition 2.6. Let P be a property that a point x in a measure space $(X, \mathcal{A}, \mu)$ may have. We say that P happens *almost everywhere* if $B = \{x \in X : P(x) \text{ is false}\}$ is in $\mathcal{A}$ and $\mu(B) = 0$.

Now, it's nice that so many functions are Riemann-integrable. But some functions aren't, because they are discontinuous on a set of positive measure. Yet, sometimes it's still "obvious" what their integral is supposed to be.

Example 2.7. Define $f : \mathbb{R} \to \mathbb{R}$ by

$$f(x) = \begin{cases} 1 & x \in \mathbb{Q}, \\ 0 & x \notin \mathbb{Q}. \end{cases}$$

Then f is not integrable on $[0, 1]$, because it is discontinuous everywhere. Yet, there are rather few points $x \in [0, 1]$ with $f(x) = 1$, so f seems like it's essentially the 0 function. We expect that $\int_0^1 f(x)\,dx$ should be 0, if only we had a better definition of the integral.

2.3 The Lebesgue Integral

Measure theory was built exactly to handle the problem of this last example: we have a function that isn't integrable but should be. So, we need to find a new definition that makes this function integrable. Measure theory is exactly the right setup for doing this. We build up to the Lebesgue integral in steps, first by explaining how to integrate a particularly nice set of functions known as *simple functions*, and then explaining how to approximate a more general class of functions by means of simple functions.

Definition 2.8. Let $(X, \mathcal{A}, \mu)$ be a measure space. For $A \in \mathcal{A}$, define the *indicator function* of A to be the function

$$\mathbb{1}_A(x) = \begin{cases} 1 & x \in A, \\ 0 & x \notin A. \end{cases}$$

Now we define the integral of an indicator function in a very straightforward way:

Definition 2.9. Let $(X, \mathcal{A}, \mu)$ be a measure space. For $A \in \mathcal{A}$, define the *(Lebesgue) integral* of $\mathbb{1}_A$ to be

$$\int_X \mathbb{1}_A \, d\mu = \mu(A).$$

In fact, this is already enough to resolve the issue in the example above: the function f is just $\mathbb{1}_{\mathbb{Q}}$, so if $X = [0,1]$ and $\mathcal{A}$ is the restriction of the Lebesgue σ-algebra on $\mathbb{R}$ to $[0,1]$, then we have $\int_{[0,1]} \mathbb{1}_{\mathbb{Q}} \, d\lambda = 0$, as expected.

But of course, not all functions are indicator functions, so we have to develop a theory for integrating other sorts of functions. Next up are simple functions.

Definition 2.10. Let $(X, \mathcal{A}, \mu)$ be a measure space. A *simple function* is a function $f : X \to \mathbb{R}$ such that there exist finitely many measurable subsets $A_1, A_2, \dots, A_n \in \mathcal{A}$ and real numbers $c_1, \dots, c_n$ such that

$$f(x) = \sum_{i=1}^{n} c_i \mathbb{1}_{A_i}(x).$$

In particular, a simple function only takes on finitely many values. We can, if we so desire, choose the A_i's to be disjoint, so that the only values taken on by f are the c_i's.

Next, we need to explain how to integrate simple functions. This is also easy:

Definition 2.11. Let $(X, \mathcal{A}, \mu)$ be a measure space, and let $f(x) = \sum_{i=1}^{n} c_i \mathbb{1}_{A_i}(x)$ be a simple function. Then we define the integral of f to be

$$\int_X f \, d\mu = \sum_{i=1}^{n} c_i \mu(A_i).$$

This integral has various good properties we expect of integrals or areas. For example, if f and g are simple functions, and $\alpha, \beta \in \mathbb{R}$, then we have

$$\int_X (\alpha f + \beta g) \, d\mu = \alpha \int_X f \, d\mu + \beta \int_X g \, d\mu,$$

and if $f(x) \leq g(x)$ for all $x \in X$, then

$$\int_X f\,d\mu \leq \int_X g\,d\mu.$$

Now, we're ready to move on from simple functions to a general class of functions.

Definition 2.12. Let X be a set and $\mathcal{A}$ a σ-algebra on X. Let $f : X \to \mathbb{R}$ be a function. We say that f is *measurable* (or $\mathcal{A}$-measurable) if, for all $t \in \mathbb{R}$, we have $\{x \in X : f(x) \leq t\} \in \mathcal{A}$.

We could just as easily replace the condition $f(x) \leq t$ with $f(x) < t$ or $f(x) \geq t$ or $f(x) > t$, and the notions would coincide.

Of course, there are nonmeasurable functions, and it's easy to describe one given the existence of a nonmeasurable set. Let S be a nonmeasurable set. Then the indicator function $\mathbb{1}_{S^c}$ of S^c is a nonmeasurable function, because $\{x \in X : f(x) \leq 0\}$ is the nonmeasurable set S.

We may now define the Lebesgue integral of arbitrary nonnegative measurable functions.

Definition 2.13. Let $(X, \mathcal{A}, \mu)$ be a measure space, and let $f : X \to \mathbb{R}_{\geq 0}$ be a measurable function. We define its *Lebesgue integral* to be

$$\int_X f\,d\mu = \sup\left\{\int_X s\,d\mu : s \text{ is simple and } s(x) \leq f(x) \text{ for all } x\right\},$$

if this supremum is finite. When the supremum is finite, we say that f is *integrable*. We write $L^1(\mu)$ for the set of integrable functions. (More precisely, $L^1(\mu)$ consists of all measurable functions $f : X \to \mathbb{R}$ such that $|f|$ is integrable. Soon, we will explain how to define the integral for functions which take on some negative values.)

In other words, we approximate f from below by simple functions. Then the integrals of these simple functions are lower bounds for the integral of f; the integral of f is the supremum of the integrals of these lower-bounding simple functions. See Figure 2.3.

Definition 2.14. If $f \in L^1(\mu)$, we write

$$\|f\|_1 = \int_X |f|\,d\mu.$$

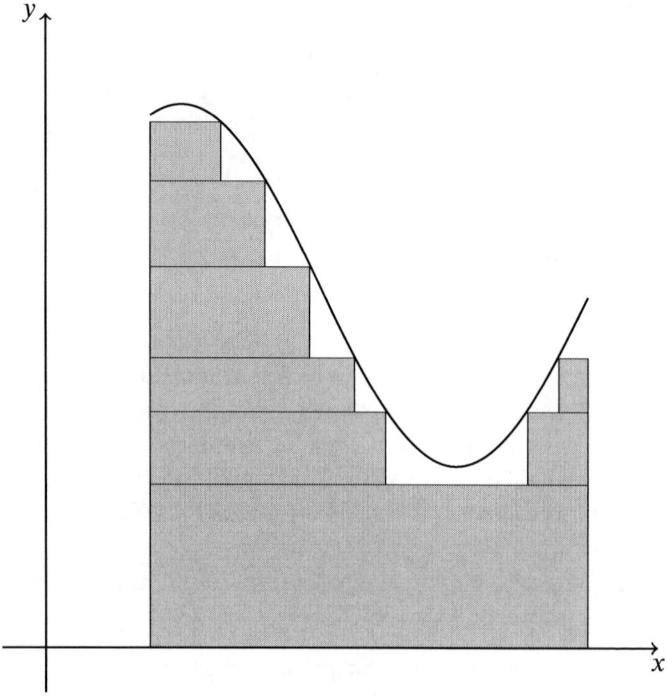

Figure 2.3. The Lebesgue integral

We call $\|f\|_1$ the L^1 *norm* of f.

Sometimes, we may wish to define an integral over a measurable subset A of X. To do that, we simply define $\int_A f\,d\mu = \int_X f \cdot \mathbb{1}_A\,d\mu$.

Measurable functions behave pretty well with respect to integrals:

Proposition 2.15. *If $(X, \mathcal{A}, \mu)$ is a measure space, $\mu(X) < \infty$, and $f : X \to \mathbb{R}_{\geq 0}$ is measurable and bounded, then f is integrable.*

The reason we focus on measurable functions for the purposes of integration is that these are the functions that are well-approximated by simple functions. We may be able to apply the definition of the Lebesgue integral in the case of the nonmeasurable function and get a number out of it, but it will have little to do with the intuitive notion of the integral as the area underneath the curve. Thus we will not be interested in assigning values to integrals of nonmeasurable functions. (However, see [**Zak66**] and [**Zak74**] for one approach to extending integration to nonmeasurable functions.)

It's worth pausing a bit to discuss the difference between the construction of the Lebesgue integral and the Riemann integral, since they are amusingly similar. For the Riemann integral of a function $f : [a, b] \to \mathbb{R}$, we chop up the domain $[a, b]$ into little pieces, see how much stuff is above each piece and sum up those contributions. With the Lebesgue integral, the situation is very similar, but this time we cut up the codomain of f (i.e. $\mathbb{R}_{\geq 0}$) into little pieces and see how much stuff is to the right (or left) of that little piece. It is surprising that this seemingly small change leads to a much more powerful theory of integration, and one that has far more desirable properties. But then perhaps it is not quite that strange, because the sets that appear in each of these horizontal strips can be far more complicated: with the Riemann integral, we're just summing up the areas of a bunch of rectangles, whereas with the Lebesgue integral, we're summing up the areas of some potentially very strange and highly disconnected sets.

Lebesgue integrals have all the same sorts of properties that Riemann integrals do, plus more. Some of the key properties enjoyed by the Riemann integral, and thus also the Lebesgue integral, are the sum rule (i.e. $\int_X (f + g)\, d\mu = \int_X f\, d\mu + \int_X f\, d\mu$) and the *Cauchy–Schwarz Inequality*

$$\left(\int_X fg\, d\mu\right)^2 \leq \left(\int_X f^2\, d\mu\right)\left(\int_X g^2\, d\mu\right)$$

whenever f, g are positive integrable functions.

Another key property is *monotonicity*:

Proposition 2.16. *Let f and g be two nonnegative integrable functions on $(X, \mathcal{A}, \mu)$, such that $f(x) \leq g(x)$ for all $x \in X$. Then*

$$\int_X f\, d\mu \leq \int_X g\, d\mu.$$

Proof. Recall the definition of $\int_X f\, d\mu$ as the supremum of the integrals of the simple functions s such that $s(x) \leq f(x)$ for all $x \in X$. If $f \leq g$, then whenever s is a simple function with $s \leq f$, then $s \leq g$ as well. Thus, if we let S be the set of integrals of simple functions $\leq f$ and T be the set of integrals of simple functions $\leq g$, then $S \subseteq T$, so $\sup(S) \leq \sup(T)$. Thus $\int_X f\,d\mu \leq \int_X g\,d\mu$. ∎

We can also integrate functions that aren't always nonnegative, but rather go from X to $\mathbb{R}$, using a little trick. Suppose $f : X \to \mathbb{R}$ is a measurable function. Then we break f up into a difference of two functions, $f = f^+ - f^-$, defined as follows:

$$f^+(x) = \max(f(x), 0), \qquad f^-(x) = \max(0, -f(x)).$$

So, for instance, if $f(x) = 4$, then $f^+(x) = 4$ and $f^-(x) = 0$, whereas if $f(x) = -4$, then $f^+(x) = 0$ and $f^-(x) = 4$. Now, if we have any measurable function $f : X \to \mathbb{R}$, we simply define the integral of f to be

$$\int_X f\,d\mu = \int_X f^+\,d\mu - \int_X f^-\,d\mu.$$

2.4 New Measures from Integration

At times, we will find it necessary to work with measures on $\mathbb{R}$ other than the Lebesgue measure. Some of the most important measures comes from the following construction. Let's work on $[0,1]$ rather than $\mathbb{R}$ for convenience, although not much will change; it's just a little bit nicer when measures are finite (i.e. $\mu(X) < \infty$).

Definition 2.17. A measure μ on X is said to be *finite* if $\mu(X) < \infty$. A measure is said to be a *probability measure* if $\mu(X) = 1$. If μ is a probability measure on $(X, \mathcal{A})$, then we say that $(X, \mathcal{A}, \mu)$ is a *probability space.*

So, $([0,1], \mathcal{L}, \lambda)$ is a probability space. We can get more of them on $[0,1]$. Take a Lebesgue integrable function f on $[0,1]$, so that $\int_{[0,1]} f\,d\lambda$ exists and is finite, and so that $f(x) \geq 0$ for all $x \in [0,1]$. Then we can define a new measure ν on $([0,1], \mathcal{L})$ by setting

$$\nu(A) = \int_A f\,d\lambda \tag{2.1}$$

for every $A \in \mathcal{L}$. We claim that ν is a measure, and even a *finite* measure. In fact, if $\int_{[0,1]} f\,d\lambda = 1$, then ν is even a probability measure. We save these verifications for Problem 4.

This is a rich source of new measures. For one important example in ergodic theory, coming from the theory of continued fractions, we will need to replace the Lebesgue measure λ with the Gauß[1] measure μ. The Gauß measure is an example of the above construction, with $f(x) = \frac{1}{\log(2)} \frac{1}{x+1}$. Thus, for all Lebesgue-measurable sets $A \subseteq [0,1]$, we have $\mu(A) = \frac{1}{\log(2)} \int_A \frac{d\lambda}{x+1}$. We shall see later why this is a good choice of measure for studying continued fractions later, when we look at invariant measures in Chapter 7.

Notice that if ν is constructed from λ in this way, then whenever $\lambda(A) = 0$, we also have $\nu(A) = 0$. It could be the case that there are some sets with $\lambda(A) \neq 0$ but $\nu(A) = 0$. However, we can avoid this situation by requiring that f be strictly positive everywhere (or almost everywhere). This will be important later on, since we tend to treat null sets very differently from sets of positive measure: we ignore null sets. So, if f is always strictly positive, then if we're interested in sets of positive measure, we don't have to distinguish between λ-measure and ν-measure.

Definition 2.18. Let $(X, \mathcal{A})$ be a σ-algebra, and let μ and ν be two measures on $(X, \mathcal{A})$. If $\nu(A) = 0$ whenever $\mu(A) = 0$, we say that ν is *absolutely continuous* with respect to μ, and we write $\nu \ll \mu$.

As we have seen, if ν is defined from λ by means of (2.1) (or with an arbitrary measure μ in place of λ) for some measurable function f, then $\nu \ll \lambda$. It turns out that there is a converse, which is one of the most important theorems in measure theory.

Theorem 2.19 (Radon–Nikodym). *Let μ and ν be two σ-finite measures*[2] *on $(X, \mathcal{A})$, with $\nu \ll \mu$. Then there exists a measurable function f such that $\nu(A) = \int_A f \, d\mu$ for all $A \in \mathcal{A}$.*

The function f is called the *Radon–Nikodym derivative* and is denoted $\frac{d\nu}{d\mu}$. See [**Rud87**, Theorem 6.10, pp. 121–123] for a proof.

---

[1]The symbol "ß" is an *Eszett*. It is a German letter (or, more precisely, ligature) for a double-s under certain circumstances. People less pretentious than I will write "Gauss" rather than "Gauß."

[2]This means that X is a countable union of sets with finite measure.

2.5 Problems

(1) Let $f : \mathbb{R} \to \mathbb{R}$ be a function. Evaluate the integral $\int_{\mathbb{R}} f \, d\delta_0$, where δ_0 is the point mass at 0. (Hint: First do this when f is an indicator function, then when f is a simple function, and finally when f is an arbitrary function.)

(2) Let X be a finite set and $\mathcal{A}$ the power set of X, i.e. the set of all subsets of X. Let μ be the counting measure on X, i.e. $\mu(A) = \#A$ for $A \subseteq X$. For a function $f : X \to \mathbb{R}$, evaluate the integral $\int_X f \, d\mu$ concretely in terms of f.

(3) Recall the definition of the Cantor set from Chapter 1. It is formed by starting with the interval $C_0 = [0, 1]$, then iteratively removing the open middle third from each interval, so that $C_1 = \left[0, \frac{1}{3}\right] \cup \left[\frac{2}{3}, 1\right]$, $C_2 = \left[0, \frac{1}{9}\right] \cup \left[\frac{2}{9}, \frac{1}{3}\right] \cup \left[\frac{2}{3}, \frac{7}{9}\right] \cup \left[\frac{8}{9}, 1\right]$, and so forth. Prove that the Cantor set is a Borel set, and that it has Lebesgue measure 0.

(4) Let f be a nonnegative function on $[0, 1]$ which is integrable with respect to the Lebesgue measure λ. For each $A \in \mathcal{L}$, define $\nu(A) = \int_A f \, d\lambda$. Show that ν is a finite measure. Show furthermore than if $\int_{[0,1]} f \, d\lambda = 1$, then ν is a probability measure.

(5) Let A be a Lebesgue measurable subset of $[0, 1]$. Prove that for any $\varepsilon > 0$, there are finitely many open intervals $(a_1, b_1), \ldots, (a_n, b_n)$ such that

$$\lambda\left(A \vartriangle \bigcup_{i=1}^{n} (a_i, b_i)\right) < \varepsilon.$$

(The notation $A \vartriangle B$ means $(A \setminus B) \cup (B \setminus A)$, i.e. the set of elements that are in exactly one of A and B.)

(6) Let $(X, \mathcal{A}, \mu)$ be a measure space and $f : X \to \mathbb{R}$ a function. Prove that if $\{x \in X : f(x) \leq t\} \in \mathcal{A}$ for all $t \in \mathbb{Q}$, then f is measurable. (Recall that the definition requires this for all $t \in \mathbb{R}$.)

(7) (a) Prove that the sum, difference, and product of two measurable functions is measurable.

(b) Prove that if f, g are measurable functions on $(X, \mathcal{A}, \mu)$, then the sets

$$\{x \in X : f(x) < g(x)\}, \qquad \{x \in X : f(x) = g(x)\}$$

are measurable.

(8) Pick an enumeration $q_1, q_2, \ldots$ of $\mathbb{Q}$ of your choosing. Cover each q_i with an interval $J_i = \left(q_i - \frac{1}{2^{i+1}}, q_i + \frac{1}{2^{i+1}}\right)$. Find infinitely many real numbers in $[0,1]$ that are not in $\bigcup_{i=1}^{\infty} J_i$. (Your answer will of course depend on your choice of enumeration.)

(9) Let $(X, \mathcal{A})$ be a σ-algebra, and let S be a countable nonempty set of nonnegative measurable functions on $(X, \mathcal{A})$. Prove that the function f, defined by $f(x) = \inf_{g \in S} g(x)$, is measurable.

(10) Prove *Chebyshev's Inequality*, which says that if $(X, \mathcal{A}, \mu)$ is a measure space and $f : X \to \mathbb{R}$ is a measurable function, then for any $t > 0$,

$$\mu(\{x \in X : |f(x)| \geq t\}) \leq \frac{1}{t} \int_X |f| \, d\mu.$$

(11) Prove that if $f : \mathbb{R} \to \mathbb{R}$ is a strictly increasing function (i.e. if $x < y$, then $f(x) < f(y)$) and $B \subseteq \mathbb{R}$ is a Borel set, then $f(B)$ is also Borel.

(12) Define a function $c : [0,1] \to [0,1]$ (the Cantor staircase function) as follows, based on the construction of the Cantor set. We define $c(x) = \frac{1}{2}$ for $\frac{1}{3} \leq x \leq \frac{2}{3}$. Then define $c(x) = \frac{1}{4}$ for $\frac{1}{9} \leq x \leq \frac{2}{9}$ and $c(x) = \frac{3}{4}$ for $\frac{7}{9} \leq x \leq \frac{8}{9}$. Continue in this way, defining c to be a constant function whose denominator is 2^n on each interval we remove in the nth step of the construction of the Cantor set, subject to the constraint that c is increasing, and no two removed intervals are assigned the same value.

(a) Prove that there is a unique way of extending c to a continuous function on all of $[0,1]$, and that this function is (weakly) increasing, with $c(0) = 0$ and $c(1) = 1$.

(b) Define $c_1 : [0,1] \to [0,1]$ by writing a number $x \in [0,1]$ in base-3. If the base-3 expansion of x contains a 1, then truncate after the first 1, then turn all the 2's into 1's and read off in binary. The resulting number is $c_1(x)$. If the base-3 expansion of x does not contain a 1, then just turn all the 2's into 1's and read off in binary. Prove that $c(x) = c_1(x)$ for all $x \in [0,1]$, so this gives you another way of defining the Cantor staircase function.

(c) Prove that if x is not in the Cantor set, then $c'(x) = 0$. Thus we have a function which is continuous and whose derivative is zero almost everywhere, but it is not a constant function!

(13) In this exercise, we will use the Cantor staircase function from the previous problem to construct a Lebesgue set that is not a Borel set. First, define $f : [0,1] \to [0,2]$ by $f(x) = c(x) + x$.

(a) Prove that f is continuous and strictly increasing. Conclude that f has an inverse function $f^{-1} : [0,2] \to [0,1]$.

(b) Prove that if I is an interval removed in the construction of the Cantor set (such as $\left(\frac{1}{3}, \frac{2}{3}\right)$ or $\left(\frac{7}{9}, \frac{8}{9}\right)$), then $f(I)$ is an interval of the same length.

(c) Prove that if C is the Cantor set in $[0,1]$, then $\lambda(f(C)) = 1$.

(d) Prove that every Lebesgue-measurable set with positive measure contains a nonmeasurable subset. (Hint: Recall the proof of the existence of a nonmeasurable set from Chapter 1.)

(e) Let N be a nonmeasurable subset of $f(C)$. Prove that $f^{-1}(N)$ is Lebesgue but not Borel. (Hint: You may find the result of Problem 11 useful.)

(f) Prove that $f^{-1}(N)$ has Lebesgue measure 0.

(14) Prove that if A is Lebesgue-measurable and $\lambda(A) > 0$, then there is a closed set $K \subseteq A$ such that $\lambda(K) > 0$. (A set is said to be *closed* if its complement is open.) In fact, prove that for any $\varepsilon > 0$, there is a closed set $K \subseteq A$ such that $\lambda(A \setminus K) < \varepsilon$.

(15) In this problem, you will prove that if $\lambda(A) > 0$, then $A - A = \{x - y : x, y \in A\}$ contains an interval of positive length.

(a) Use Problem 14 to reduce to the case where A is closed, and that A is *bounded* in the sense that there is some M such that

$|x| \leq M$ for all $x \in A$. From now on, we will assume that A is closed and bounded, also known as *compact*.

(b) Prove that there is an open set $U \supseteq A$ such that $\lambda(U) < 2\lambda(A)$.

(c) Show that there is some $\varepsilon > 0$ such that if $x \in A$ and $y \in U^c$, then $|x - y| > \varepsilon$. Note that this value of ε has to work for all $x \in A$ and $y \in U^c$.

(d) Prove that there is some $\varepsilon > 0$ such that if $|x| < \varepsilon$, then $x + A \subseteq U$. (Here $x + A = \{x + y : y \in A\}$.)

(e) Prove that $A - A$ contains the interval $(-\varepsilon, \varepsilon)$.

(16) Find a set A with $\lambda(A) = 0$ such that $A - A$ contains an interval of positive length.

Chapter 3

Some Limit Theorems

3.1 Riemann Integrals Misbehaving

The Riemann integral, while somewhat easier to define than the Lebesgue integral, does not have very good properties. Let's see some of the properties we wish an integral would have, and why the Riemann integral doesn't satisfy them.

The Riemann integral does not behave well under *monotone sequences*. What we mean by this is the following: Suppose we have a sequence of functions $f_1, f_2, \ldots$, say on $[0,1]$ but this isn't very important, such that $f_n(x) \geq 0$ for all $x \in [0,1]$, and that $f_1(x) \leq f_2(x) \leq \cdots$ for all $x \in [0,1]$. Furthermore, suppose that each f_n is integrable, i.e. $\int_0^1 f_n(x)\,dx$ exists. Let $f(x) = \lim_{n\to\infty} f_n(x)$. (This might be ∞, but the limit exists if we allow ∞ as a possible limit, since the f_n's are increasing in n.) It would be nice if $\int_0^1 f(x)\,dx$ would exist and be equal to $\lim_{n\to\infty} \int_0^1 f_n(x)\,dx$.

Unfortunately, this is not true of the Riemann integral. To see this, we'll construct a counterexample. We have already seen an example of a function that is not Riemann integrable, namely the function f such that $f(x) = 1$ if $x \in \mathbb{Q}$ and $f(x) = 0$ if $x \notin \mathbb{Q}$. So, let's find a monotone sequence that converges to f pointwise, i.e. $f(x) = \lim_{n\to\infty} f_n(x)$ for all $x \in [0,1]$. To do this, let $q_1, q_2, \ldots$ be an enumeration of the rationals in $[0,1]$. Define $f_n(x)$ by

$$f_n(x) = \begin{cases} 1 & x \in \{q_1, q_2, \ldots, q_n\}, \\ 0 & \text{otherwise.} \end{cases}$$

For every $q \in [0,1] \cap \mathbb{Q}$, there is some N such that for all $n \geq N$, we have $f_n(q) = 1$, whereas for every $r \in [0,1] \setminus \mathbb{Q}$, $f_n(r) = 0$ for all n. Thus $f(x) = \lim_{n\to\infty} f_n(x)$ for all $x \in [0,1]$.

Now, f_n is Riemann-integrable for all n, because it's continuous except on the finite set of points $\{q_1, \ldots, q_n\}$. In fact, its integral is 0, although we don't need this fact at the moment. However, as we have seen, the limit function f is not integrable on $[0,1]$. So we see that the property of being Riemann integrable is not closed under monotone limits.

Ideally, we'd really like the following. Suppose that f_n is a sequence of functions (say on $[0,1]$), all of which are integrable, and f_n converges to f pointwise in the sense that $\lim_{n\to\infty} f_n(x) = f(x)$ for all x. Then f is integrable, and $\lim_{n\to\infty} \int_0^1 f_n(x)\,dx = \int_0^1 f(x)\,dx$. In other words, we'd like to be able to switch a limit with an integral:

$$\lim_{n\to\infty} \int_0^1 f_n(x)\,dx = \int_0^1 \lim_{n\to\infty} f_n(x)\,dx.$$

This sort of thing certainly isn't possible in complete generality, as the following example shows:

Example 3.1. Let $f_n : [0,1] \to \mathbb{R}$ be defined by

$$f_n(x) = \begin{cases} n & x \in (0, \frac{1}{n}), \\ 0 & \text{otherwise.} \end{cases}$$

For every point $x \in [0,1]$, $\lim_{n\to\infty} f_n(x) = 0$, so f_n converges to 0 pointwise. However, $\int_0^1 f_n(x)\,dx = 1$ for all n.

But that happens because the f_n's aren't terribly well-behaved. For example, they aren't uniformly bounded: there isn't any number B such that $f_n(x) \leq B$ for all n and x. They aren't monotone: sometimes $f_n(x) < f_{n+1}(x)$, but sometimes $f_n(x) > f_{n+1}(x)$. So these functions are just a mess, and we can't hope for anything nice to happen with them. As we shall see, there is a patch to this for the Lebesgue integral, called *Fatou's Lemma*.

3.2 The Big Three

There are three main theorems about switching limits and integrals that hold for the Lebesgue integral. These are known as *Fatou's Lemma*, the *Monotone Convergence Theorem*, and the *Dominated Convergence Theorem*. Once we have proven any one of them, it's quite quick to prove the other two, but we have to do a bit of work to prove the first one (but it isn't too bad). Our approach will be to prove the Monotone Convergence Theorem from scratch and use this one to derive the other two, but we could just as easily make a different decision as to which one to start with, and it's only a matter of personal preference.

Definition 3.2. Let $a_1, a_2, \ldots$ be a sequence of real numbers. The *limit inferior* of the a_i's is defined to be the largest real number a such that, for every $\varepsilon > 0$, there are only finitely many a_i's less than $a - \varepsilon$. If there is no a such that there are only finitely many a_i's less than a, then the limit inferior is $-\infty$, while if for every a there are only finitely many a_i's less than a, then the limit inferior is $+\infty$. We write $\liminf_{n\to\infty} a_n$ for this number (or $-\infty$).

Another interpretation of the limit inferior is that it is the smallest possible limit of a subsequence of the a_i's. There is a dual notion, the *limit superior*, which is the largest possible limit of a subsequence. We write $\limsup_{n\to\infty} a_n$ for that. In the case that $\lim_{n\to\infty} a_n$ exists, then $\lim_{n\to\infty} a_n = \liminf_{n\to\infty} a_n = \limsup_{n\to\infty} a_n$.

Example 3.3. Consider the sequence $a_n = (-1)^n\left(1 + \frac{1}{n}\right)$, so that the sequence goes $-2, \frac{3}{2}, -\frac{4}{3}, \frac{5}{4}, -\frac{6}{5}, \ldots$ Then $\liminf_{n\to\infty} a_n = -1$ and $\limsup_{n\to\infty} a_n = 1$.

We can also take limits inferior of functions. If we have functions f_n, then the limit inferior is the function

$$f(x) = \liminf_{n\to\infty} f_n(x),$$

i.e. the pointwise limit inferior.

On the other hand, we also have the non-limit versions, called *infima* and *suprema*.

Definition 3.4. Let A be a sequence of numbers. The *infimum* of A is the largest real number s such that $s \leq a$ for all $a \in A$. We write $\inf A$ for this number. Similarly, the *supremum* is the smallest real number t such that $t \geq a$ for all $a \in A$. We write $\sup A$ for the supremum.

The infimum and supremum are the infinite versions of minima and maxima. We can't necessarily find the minimum of an infinite set, because the minimum might not be attained. For example, if $A = \{\frac{1}{n} : n = 1, 2, \ldots\}$, then A does not have a minimum, but it has an infimum, namely 0.

Theorem 3.5 (Fatou's Lemma). *Let $(X, \mathcal{A}, \mu)$ be a measure space, and let $f_1, f_2, \ldots$ be a sequence of measurable functions from X to $[0, \infty]$. Then*

$$\int_X \liminf_{n\to\infty} f_n \, d\mu \leq \liminf_{n\to\infty} \int_X f_n \, d\mu.$$

Theorem 3.6 (Monotone Convergence Theorem). *Let $(X, \mathcal{A}, \mu)$ be a measure space. Let $f_1, f_2, \ldots$ be a sequence of measurable functions from X to $[0, \infty]$ such that for all $x \in X$, we have $0 \leq f_1(x) \leq f_2(x) \leq \cdots$. Let $f(x) = \lim_{n\to\infty} f_n(x)$. Then*

$$\int_X f \, d\mu = \lim_{n\to\infty} \int_X f_n \, d\mu.$$

Theorem 3.7 (Dominated Convergence Theorem). *Let $(X, \mathcal{A}, \mu)$ be a measure space. Let $g : X \to [0, \infty]$ be an integrable function, and let $f_1, f_2, \ldots$ be measurable functions from X to $[0, \infty]$ be such that for each $x \in X$, $\lim_{n\to\infty} f_n(x)$ exists. Call this limit function f:*

$$f(x) = \lim_{n\to\infty} f_n(x).$$

Suppose furthermore that $f_n(x) \leq g(x)$ for all n and x. Then f and all the f_n's are integrable, and

$$\int_X f \, d\mu = \lim_{n\to\infty} \int_X f_n \, d\mu.$$

A natural question to ask about is the inequality in Fatou's Lemma. Why don't we have $\int_X \liminf_{n\to\infty} f_n \, d\mu = \liminf_{n\to\infty} \int_X f_n \, d\mu$? We have already seen the relevant counterexample: the functions

$$f_n(x) = \begin{cases} n & x \in \left(0, \frac{1}{n}\right), \\ 0 & \text{otherwise.} \end{cases}$$

Then we have $\liminf_{n\to\infty} f_n(x) = 0$ for all x so

$$\int_{[0,1]} \liminf_{n\to\infty} f_n \, d\lambda = 0 \qquad \text{but} \qquad \liminf_{n\to\infty} \int_{[0,1]} f_n \, d\lambda = 1.$$

So we really do need an inequality there.

Okay, now let's turn to the proofs! As promised, we'll start with a proof of the Monotone Convergence Theorem, and then we'll deduce Fatou's Lemma and the Dominated Convergence Theorem as corollaries.

Proof of the Monotone Convergence Theorem. We make no assumptions about the integrability of the f_n's, so we might have $\int_X f_n \, d\mu = \infty$ for some n, and that's fine. By the monotonicity of the Lebesgue integral, we have

$$\int_X f_1 \, d\mu \le \int_X f_2 \, d\mu \le \cdots \le \int_X f \, d\mu.$$

Thus the sequence of the numbers or ∞ $(\int_X f_n \, d\mu)$ converges to something, possibly ∞, and we want to show that the limit is $\int_X f \, d\mu$. Since $\int_X f_n \, d\mu \le \int_X f \, d\mu$ for all n, we have $\lim_{n\to\infty} \int_X f_n \, d\mu \le \int_X f \, d\mu$. So, it suffices to prove that $\int_X f \, d\mu \le \lim_{n\to\infty} \int_X f_n \, d\mu$. For this we need to recall the definition of the integral, as a limit of integrals of simple functions.

For each n, choose a nondecreasing sequence $g_{n,1}, g_{n,2}, \ldots$ of simple measurable functions such that $g_{n,k}(x) \le g_{n,k+1}(x)$ for all k and x, and where each $g_{n,k} : X \to [0, \infty)$, and such that $\lim_{k\to\infty} g_{n,k}(x) = f_n(x)$ for all x. Now we diagonalize: define h_n by

$$h_n(x) = \max(g_{1,n}(x), g_{2,n}(x), \ldots, g_{n,n}(x)).$$

The h_n's are all simple functions, they satisfy $h_n(x) \le f_n(x)$ for all x and n, and $\lim_{n\to\infty} h_n(x) = f(x)$ for all x. Then by something quite close to the definition of the integral, we have $\int_X f \, d\mu = \lim_{n\to\infty} \int_X h_n \, d\mu$. By monotonicity again, we have

$$\int_X f\,d\mu = \lim_{n\to\infty}\int_X h_n\,d\mu \le \lim_{n\to\infty}\int_X f_n\,d\mu,$$

proving the other inequality. Thus $\int_X f\,d\mu = \lim_{n\to\infty}\int_X f_n\,d\mu$, as desired. ■

Proof of Fatou's Lemma. We extract a monotone sequence out of the f_n's. We let

$$g_n(x) = \inf_{k\ge n} f_k(x).$$

Each g_n is measurable because the infimum of measurable functions is always measurable, and we have $g_1(x) \le g_2(x) \le \cdots$ for all x. Furthermore, $\liminf f_n(x) = \lim_{n\to\infty} g_n(x)$ (see Problem 4). By the Monotone Convergence Theorem, then, we have

$$\begin{aligned}\int_X \liminf_{n\to\infty} f_n\,d\mu &= \int_X \lim_{n\to\infty} g_n\,d\mu\\ &= \lim_{n\to\infty}\int_X g_n\,d\mu\\ &= \lim_{n\to\infty}\int_X \inf_{k\ge n} f_k\,d\mu\\ &\le \liminf_{n\to\infty}\int_X f_n\,d\mu,\end{aligned}$$

as desired. ■

There's a useful "reverse" version of Fatou's Lemma that we will need next, in the proof of the Dominated Convergence Theorem.

Lemma 3.8 (Reverse Fatou's Lemma). *Let $(X,\mathcal{A},\mu)$ be a measure space, and let $f_1, f_2, \ldots$ be a sequence of measurable functions from X to $[0,\infty]$, such that there exists an integrable function g such that $f_n \le g$ for all n. Then*

$$\limsup_{n\to\infty}\int_X f_n\,d\mu \le \int_X \limsup_{n\to\infty} f_n\,d\mu.$$

Proof. We apply Fatou's Lemma to the functions $g - f_n$, so that we have

$$\int_X \liminf_{n\to\infty}(g-f_n)\,d\mu \le \liminf_{n\to\infty}\int_X (g-f_n)\,d\mu.$$

Since $\liminf_{n\to\infty}(g - f_n) = g - \limsup_{n\to\infty} f_n$, we have

$$\int_X g\,d\mu - \int_X \limsup_{n\to\infty} f_n\,d\mu \le \int_X g\,d\mu - \limsup_{n\to\infty} \int_X f_n\,d\mu,$$

or

$$\limsup_{n\to\infty} \int_X f_n\,d\mu \le \int_X \limsup_{n\to\infty} f_n\,d\mu,$$

as desired. ∎

Proof of the Dominated Convergence Theorem. The integrability of the f_n's follows from the monotonicity of the integral: since g is integrable and $f_n \le g$, we have $\int_X f_n\,d\mu \le \int_X g\,d\mu < \infty$, so f_n is integrable. To show that $\int_X f\,d\mu = \lim_{n\to\infty} \int_X f_n\,d\mu$, we'll actually show something even stronger, namely that

$$\lim_{n\to\infty} \int_X |f_n - f|\,d\mu = 0.$$

This implies the result by the *triangle inequality*, which says that

$$\int_X |h|\,d\mu \ge \left|\int_X h\,d\mu\right|.$$

Applying this to $h = f_n - f$, we get

$$\lim_{n\to\infty} \int_X |f_n - f|\,d\mu \ge \lim_{n\to\infty} \left|\int_X (f_n - f)\,d\mu\right|,$$

so the right side is 0 because the left side is.

Now, note that $|f - f_n| \le f + f_n \le 2g$ for all n, and that

$$\limsup_{n\to\infty} |f(x) - f_n(x)| = 0$$

for all $x \in X$. Now, we have

$$\left|\int_X f\,d\mu - \int_X f_n\,d\mu\right| = \left|\int_X (f - f_n)\,d\mu\right| \le \int_X |f - f_n|\,d\mu.$$

By the Reverse Fatou's Lemma, we have

$$\limsup_{n\to\infty} \int_X |f - f_n|\,d\mu \le \int_X \limsup_{n\to\infty} |f - f_n|\,d\mu = 0.$$

Thus the limsup on the left is actually a limit, and it's equal to 0, i.e.

$$\lim_{n\to\infty} \int_X |f - f_n|\,d\mu = 0.$$

∎

3.3 An Application

Let's see a trick we can do with our big theorems.

Proposition 3.9. *Let $(X, \mathcal{A}, \mu)$ be a measure space, and suppose $f_1, f_2, \ldots$ is a sequence of measurable functions in $L^1(\mu)$ such that f_n converges to a function f almost everywhere, i.e. for almost all $x \in X$, $\lim_{n\to\infty} f_n(x) = f(x)$. Suppose also that $\lim_{n\to\infty} \|f_n\|_1 = \|f\|_1$. Then for all $A \in \mathcal{A}$, we have*

$$\lim_{n\to\infty} \int_A |f_n|\, d\mu = \int_A |f|\, d\mu.$$

Proof. Note that we have

$$|f_n - f|\mathbb{1}_A \le |f_n - f| \le |f_n| + |f|.$$

Let $g_n = |f_n| + |f| - |f_n - f|\mathbb{1}_A$, which is a nonnegative function, and apply Fatou's Lemma. This gives

$$\int_X \liminf_{n\to\infty} g_n\, d\mu \le \liminf_{n\to\infty} \int_X g_n\, d\mu,$$

or

$$\int_X \liminf_{n\to\infty}(|f_n| + |f| - |f_n - f|\mathbb{1}_A)\, d\mu$$
$$\le \liminf_{n\to\infty} \int_X (|f_n| + |f| - |f_n - f|\mathbb{1}_A)\, d\mu,$$

or

$$\int_X (2|f| - \limsup_{n\to\infty} |f_n - f|\mathbb{1}_A)\, d\mu$$
$$\le \liminf_{n\to\infty} \int_X |f_n|\, d\mu + \|f\|_1 - \limsup_{n\to\infty} \int_A |f_n - f|\, d\mu,$$

or

$$2\|f\|_1 \le 2\|f\|_1 - \limsup_{n\to\infty} \int_A |f_n - f|\, d\mu,$$

since $\limsup_{n\to\infty} |f_n - f| = 0$ almost everywhere. Thus we have

$$\limsup_{n\to\infty} \int_A |f_n - f|\, d\mu \le 0.$$

Furthermore, since an integral of a nonnegative function is nonnegative, we have

$$0 \leq \liminf_{n\to\infty} \int_A |f_n - f|\, d\mu \leq \limsup_{n\to\infty} \int_A |f_n - f|\, d\mu \leq 0,$$

so $\lim \int_A |f_n - f|\, d\mu = 0$. Now, by the triangle inequality, we have

$$\begin{aligned}
\lim_{n\to\infty} \left| \int_A |f_n|\, d\mu - \int_A |f|\, d\mu \right| &= \left| \int_A (|f_n| - |f|)\, d\mu \right| \\
&\leq \lim_{n\to\infty} \int_A ||f_n| - |f||\, d\mu \\
&\leq \lim_{n\to\infty} \int_A |f_n - f|\, d\mu \\
&= 0,
\end{aligned}$$

as desired. ∎

This is often the sort of way in which Fatou's Lemma and the others are used in practice.

Using some of these big limit theorems, especially the Monotone Convergence Theorem and the Dominated Convergence Theorem, is how we often compute Lebesgue integrals. Unlike the Riemann integral, we don't have a database of techniques for evaluating them. However, if we have a function that is Riemann integrable, it is also Lebesgue integrable, and the integrals are equal. But if we have a function f which is Lebesgue integrable and not Riemann integrable, we might be able to write it as a limit of Riemann integrable functions f_n, and then use one of the limit theorems to evaluate $\int_X f\, d\mu$ as $\lim_{n\to\infty} \int_X f_n\, d\mu$, assuming the hypotheses apply.

3.4 Types of Convergence

Given a sequence of measurable functions $f_1, f_2, \ldots$ on a measure space $(X, \mathcal{A}, \mu)$, there are several senses in which they can converge to another measurable function f.

Definition 3.10. We say that f_n converges to f *pointwise* if, for all $x \in X$, $\lim_{n\to\infty} f_n(x) = f(x)$.

Definition 3.11. We say that f_n converges to f *almost everywhere* if there is a null set $N \subseteq X$ such that for all $x \in X \setminus N$, $\lim_{n\to\infty} f_n(x) = f(x)$.

Definition 3.12. We say that f_n converges to f *in measure* if, for every $\varepsilon > 0$, we have

$$\lim_{n\to\infty} \mu(\{x \in X : |f_n(x) - f(x)| > \varepsilon\}) = 0.$$

In general, there is no relation between convergence almost everywhere and convergence in measure. To see this, consider the following two examples, both on $(\mathbb{R}, \mathcal{L}, \lambda)$.

Example 3.13. Let $f_n = \mathbb{1}_{[n,\infty)}$. Then f_n converges to 0 almost everywhere (in fact, everywhere!), but f_n does *not* converge to 0 in measure, because $\mu(\{x : f_n(x) = 1\}) = \infty$ for all n.

Example 3.14. Consider the sequence f_n, where $f_1 = \mathbb{1}_{[0,1]}$, $f_2 = \mathbb{1}_{\left[0,\frac{1}{2}\right]}$, $f_3 = \mathbb{1}_{\left[\frac{1}{2},1\right]}$, $f_4 = \mathbb{1}_{\left[0,\frac{1}{4}\right]}$, $f_5 = \mathbb{1}_{\left[\frac{1}{4},\frac{1}{2}\right]}$, $f_6 = \mathbb{1}_{\left[\frac{1}{2},\frac{3}{4}\right]}$, $f_7 = \mathbb{1}_{\left[\frac{3}{4},1\right]}$, and so forth: these functions are characteristic functions of intervals of length $\frac{1}{2^k}$, where have 2^k such intervals covering $[0,1]$. Then f_n converges to 0 in measure, but not almost everywhere: for any $x \in [0,1]$, there are infinitely many values of n such that $f_n(x) = 1$.

Of these, the second example seems like the more serious of the two, and indeed, it is.

Proposition 3.15. *Let $(X, \mathcal{A}, \mu)$ be a finite measure space, and let $f, f_1, f_2, \ldots$ be measurable functions on X. If f_n converges to f almost everywhere, then f_n converges to f in measure.*

Hence the need to work with all of $\mathbb{R}$, or at least some space with infinite measure, in the first example.

Proof. We need to show that for all $\varepsilon > 0$, we have

$$\lim_{n\to\infty} \mu(\{x \in X : |f_n(x) - f(x)| > \varepsilon\}) = 0.$$

Pick an $\varepsilon > 0$, and define sets $A_1, A_2, \ldots, B_1, B_2, \ldots$ by

$$A_n = \{x \in X : |f_n(x) - f(x)| > \varepsilon\}, \qquad B_n = \bigcup_{k=n}^{\infty} A_k.$$

The sequence B_n is decreasing, in the sense that $B_n \supseteq B_{n+1}$ for all n. Furthermore, if $x \in \bigcap_{n=1}^{\infty} B_n$, then $f_n(x)$ does not converge to $f(x)$. Since f_n converges to f almost everywhere, this means that $\mu\left(\bigcap_{n=1}^{\infty} B_n\right) = 0$. Since $\mu\left(\bigcap_{n=1}^{\infty} B_n\right) = \lim_{n\to\infty} \mu(B_n)$, the latter limit is 0. Since $A_n \subseteq B_n$, we have

$$\begin{aligned}\lim_{n\to\infty} \mu(\{x \in X : |f_n(x) - f(x)| > \varepsilon\}) &= \lim_{n\to\infty} \mu(A_n)\\ &\leq \lim_{n\to\infty} \mu(B_n)\\ &= 0.\end{aligned}$$

Thus f_n converges to f in measure, as desired. ■

As we have seen, there is no converse to Proposition 3.15: convergence in measure does *not* imply convergence almost everywhere, even on a finite measure space. However, it's close:

Proposition 3.16. *Let $(X, \mathcal{A}, \mu)$ be a measure space, and let $f, f_1, f_2, \ldots$ be measurable functions on X. If f_n converges to f in measure, then there is a subsequence of f_n that converges to f almost everywhere.*

Proof. Suppose f_n converges to f in measure. Then, by definition, we have

$$\lim_{n\to\infty} \mu(\{x \in X : |f_n(x) - f(x)| > \varepsilon\}) = 0$$

for all $\varepsilon > 0$. Thus we can construct a subsequence f_{k_n} of the f_n's as follows: choose k_1 such that

$$\mu(\{x \in X : |f_{k_1}(x) - f(x)| > 1\}) \leq \frac{1}{2}.$$

Then, assuming we have already chosen k_{n-1}, choose $k_n > k_{n-1}$ such that

$$\mu\left(\left\{x \in X : |f_{k_n}(x) - f(x)| > \frac{1}{n}\right\}\right) \leq \frac{1}{2^n}.$$

Define sets $A_1, A_2, \ldots$ by

$$A_n = \left\{x \in X : |f_{k_n}(x) - f(x)| > \frac{1}{n}\right\}.$$

If $x \notin \bigcap_{j=1}^{\infty} \bigcup_{n=j}^{\infty} A_n$, then there is a positive integer j such that $x \notin \bigcup_{n=j}^{\infty} A_n$, i.e. such that $|f_{k_n}(x) - f(x)| \leq \frac{1}{n}$ for $n \geq j$, meaning that $f_{k_n}(x)$ converges to $f(x)$. Thus f_{k_n} converges to f outside $\bigcap_{j=1}^{\infty}\bigcup_{n=j}^{\infty} A_n$. Since

$$\mu\left(\bigcup_{n=j}^{\infty} A_n\right) \leq \sum_{n=j}^{\infty} \mu(A_n) \leq \sum_{n=j}^{\infty} \frac{1}{2^n} = \frac{1}{2^{j-1}}$$

for all j, we have

$$\mu\left(\bigcap_{j=1}^{\infty}\bigcup_{n=j}^{\infty} A_n\right) \leq \mu\left(\bigcup_{n=j}^{\infty} A_n\right) \leq \frac{1}{2^{j-1}}$$

for all j, so

$$\mu\left(\bigcap_{j=1}^{\infty}\bigcup_{n=j}^{\infty} A_n\right) = 0,$$

as desired. ■

3.5 L^p Spaces

Given a measure space $(X, \mathcal{A}, \mu)$, we can distinguish several types of functions. We have already seen the $\mathcal{L}^1$ functions, i.e. those measurable functions f for which $\int_X |f|\, d\mu < \infty$. We can generalize this notion a bit:

Definition 3.17. Let $1 \leq p < \infty$. The $\mathcal{L}^p$ functions with respect to $(X, \mathcal{A}, \mu)$ are those measurable functions f on X for which

$$\int_X |f|^p\, d\mu < \infty.$$

We write $\mathcal{L}^p(\mu)$ for the set of all $\mathcal{L}^p$ functions. For $f \in \mathcal{L}^p(\mu)$, we define its L^p*-norm* to be

$$\|f\|_p = \left(\int_X |f|^p\, d\mu\right)^{1/p}.$$

We can also define $\mathcal{L}^\infty$ functions and the L^∞-norm. To do this, we need the following notion:

Definition 3.18. Let $f : X \to \mathbb{R}$ be a measurable function on $(X, \mathcal{A}, \mu)$. The *essential supremum* of f is defined to be

$$\text{ess sup}(f) = \inf\{a : \mu(f^{-1}((a,\infty))) = 0\}.$$

If there is no such a, then $\text{ess sup}(f) = \infty$.

In other words, $\text{ess sup}(f)$ is the supremum up to a null set.

Definition 3.19. The L^∞ norm $\|f\|_\infty$ of f is $\text{ess sup}(|f|)$.

In Problem 8, you will confirm that this is a sensible definition.

It is often useful to consider two functions in $\mathcal{L}^p$ to be the same if they agree up to a null set. We write $L^p(\mu)$ for the set of functions in $\mathcal{L}^p(\mu)$, except that $f, g \in \mathcal{L}^p(\mu)$ are considered equal if $\mu(\{x \in X : f(x) \neq g(x)\}) = 0$. This is a bit nicer, because then we have $\|f\|_p = 0$ if and only if $f = 0$ in $L^p(\mu)$. This isn't true in $\mathcal{L}^p(\mu)$, because if $f(x) = 0$ except on a null set, then $\int_X |f|\, d\mu$ is still zero, so $\|f\|_p = 0$. Thus the elements of L^p are not exactly functions but rather equivalence classes of functions. But in practice it's still okay to think of them as being functions, and we will do so.

The L^p spaces have the following nice fundamental properties:

- $\|f\|_p = 0$ if and only if $f = 0$ in L^p.
- For all $f \in L^p(\mu)$ and $\lambda \in \mathbb{R}$, we have $\|\lambda f\|_p = |\lambda| \cdot \|f\|_p$.
- (Hölder's Inequality) Suppose $1 \le p, q \le \infty$ satisfy $\frac{1}{p} + \frac{1}{q} = 1$. If $f \in L^p(\mu)$ and $g \in L^q(\mu)$, then $fg \in L^1(\mu)$, and we have
$$\|fg\|_1 \le \|f\|_p \|g\|_q.$$
- (Minkowski's Inequality) Suppose $f, g \in L^p(\mu)$. Then $f + g \in L^p(\mu)$ as well, and we have $\|f + g\|_p \le \|f\|_p + \|g\|_p$.

The proofs of Hölder's Inequality and Minkowski's Inequality aren't trivial, but as their proofs won't be needed elsewhere, we omit them. See [**Rud87**, Theorem 3.5, pp. 63–65] for proofs of Hölder's Inequality and Minkowski's Inequality.

Another very important fact about L^p spaces is that they are *complete*. What that means is that if we have a sequence $f_1, f_2, \ldots$ of functions in

$L^p(\mu)$ that get close to each other in the L^p norm, then they converge (in a suitable sense) to another function in $L^p(\mu)$. More precisely, suppose that for every $\varepsilon > 0$, there is some positive integer N such that whenever $m, n > N$, $\|f_m - f_n\|_p < \varepsilon$. Then there exists some $f \in L^p(\mu)$ such that $\lim_{n\to\infty} \|f_n - f\|_p = 0$.

This is very nice, because it allows us to define functions as limits of sequences of functions that we understand better. If we replace the Lebesgue integral with the Riemann integral, we can't necessarily do that, because limits of L^p functions with respect to the Riemann integral can be badly discontinuous, hence not Riemann integrable.

3.6 Problems

(1) Prove the following properties of limits inferior:

(a) Let $a_1, a_2, \ldots$ and $b_1, b_2, \ldots$ be two sequences of real numbers. Then

$$\liminf_{n\to\infty}(a_n + b_n) \geq \liminf_{n\to\infty} a_n + \liminf_{n\to\infty} b_n.$$

Find an example where inequality is strict.

(b) If $a_1 \geq a_2 \geq a_3 \geq \cdots$, then $\liminf_{n\to\infty} a_n = \inf a_n$.

(c) For any sequence $a_1, a_2, a_3, \ldots$, we have

$$\liminf_{n\to\infty} a_n \leq \limsup_{n\to\infty} a_n.$$

(d) Let $a_1, a_2, \ldots$ and $b_1, b_2, \ldots$ be sequences of *nonnegative* real numbers, and suppose that $\lim_{n\to\infty} b_n = b$. Then

$$\liminf_{n\to\infty} a_n b_n = b \cdot \liminf_{n\to\infty} a_n.$$

(2) Find a sequence $f_1, f_2, \ldots$ of Lebesgue-measurable functions on $[0,1]$ such that $\lim_{n\to\infty} f_n(x) = 0$ for all $x \in [0,1]$, but

$$\lim_{n\to\infty} \int_{[0,1]} f_n \, d\lambda = \infty.$$

(3) Find a sequence $f_1, f_2, \ldots$ of Lebesgue-measurable functions on $[0,1]$ such that $\lim_{n\to\infty} f_n(x) = 0$ for all $x \in [0,1]$, but

$$\limsup_{n\to\infty} \int_{[0,1]} f_n \, d\lambda = 1 \quad \text{and} \quad \liminf_{n\to\infty} \int_{[0,1]} f_n \, d\lambda = -1.$$

(4) Let $f_1, f_2, \ldots$ be a sequence of measurable functions, and define a new sequence of functions $g_1, g_2, \ldots$ by setting

$$g_n(x) = \inf_{k \geq n} f_k(x).$$

Prove that $g_1(x) \leq g_2(x) \leq \cdots$, $\lim_{n\to\infty} g_n(x)$ exists for all x (it might be ∞), and that $\liminf f_n(x) = \lim_{n\to\infty} g_n(x)$.

(5) Prove *Beppo Levi's Theorem*, which says that if $(X, \mathcal{A}, \mu)$ is a measure space and $f_1, f_2, \ldots$ is a sequence of measurable functions from X to $[0, \infty]$, then

$$\int_X \sum_{k=1}^{\infty} f_k \, d\mu = \sum_{k=1}^{\infty} \int_X f_k \, d\mu.$$

(6) Prove that the Monotone Convergence Theorem still holds if we drop the assumption that the f_n's are nonnegative, but instead add the condition that f_1 be integrable, i.e. $\int_X f_1 \, d\mu$ exists and is finite.

(7) Evaluate

$$\lim_{n\to\infty} \int_0^T \sin(1 - e^{-x/n}) e^{-x} \, dx$$

as a function of T. Use one of the limit theorems to prove that your answer is correct.

(8) Prove that $\|f\|_\infty = \lim_{p\to\infty} \|f\|_p$.

(9) Let $(X, \mathcal{A}, \mu)$ be a measure space, and let $f : X \to [0, \infty]$ be measurable.

(a) Show that if $f(x)$ is a nonnegative integer or $+\infty$ for all $x \in X$, then

$$\int_X f \, d\mu = \sum_{n=1}^{\infty} \mu(\{x : f(x) \geq n\}).$$

(b) Suppose that $\mu(X) < \infty$. Show that f is integrable if and only if the series

$$\sum_{n=1}^{\infty} \mu(\{x : f(x) \geq n\})$$

converges.

(10) Let $(X, \mathcal{A}, \mu)$ be a measure space, and let f and $f_1, f_2, \ldots$ be nonnegative integrable functions such that $\lim_{n\to\infty} f_n(x) = f(x)$ for almost all $x \in X$, and $\lim_{n\to\infty} \int_X f_n \, d\mu = \int_X f \, d\mu$. Prove that

$$\lim_{n\to\infty} \int_X |f_n - f| \, d\mu = 0.$$

The latter condition is known as convergence *in mean.*

(11) Prove that if $f_1, f_2, \ldots$ converge to f in mean, then they converge to f in measure.

Chapter **4**

Measure-Preserving Transformations

4.1 Measure-Preserving Transformations

The main object of study in ergodic theory is that of a measure-preserving transformation. Let us now define what that means.

Definition 4.1. Let $(X, \mathcal{A}, \mu)$ be a measure space. A function $T : X \to X$ is said to be a *measure-preserving transformation* if for all $A \in \mathcal{A}$, we have $T^{-1}(A) \in \mathcal{A}$, and $\mu(T^{-1}(A)) = \mu(A)$. By $T^{-1}(A)$, we mean $\{x \in X : T(x) \in A\}$; this is the *preimage* of A.

Generally, we would like $(X, \mathcal{A}, \mu)$ to be a *finite* measure space, i.e. one in which $\mu(X) < \infty$, in which case by rescaling we may freely assume that $\mu(X) = 1$ so that X is a probability space.

Definition 4.2. A *measure-preserving system* is a quadruple $(X, \mathcal{A}, \mu, T)$, where $(X, \mathcal{A}, \mu)$ is a probability space and $T : X \to X$ is a measure-preserving transformation.

Let's see some typical examples of measure-preserving systems. These are some of the ones that we'll be studying throughout the rest of the book, as they are rather typical examples in many respects.

Example 4.3. Let $X = [0, 1)$, let $\mathcal{A}$ be the Lebesgue σ-algebra on X, and let μ be the Lebesgue measure on X. Let θ be a real number, and consider the transformation $R_\theta : X \to X$, defined by $R_\theta(x) = (x + \theta) \pmod 1$, or equivalently $R_\theta(x) = x+\theta-\lfloor x+\theta \rfloor$. Let's check that this transformation is measure-preserving. First, $R_\theta^{-1}(A) = (A-\theta) \pmod 1$, so it's pretty clear that $R_\theta^{-1}(A) \in \mathcal{A}$ whenever $A \in \mathcal{A}$. As for the measure-preserving property, if $A \in \mathcal{A}$, then $\mu(R_\theta^{-1}(A)) = \lambda(A-\theta) = \lambda(A) = \mu(A)$, where λ is the Lebesgue measure on $\mathbb{R}$. Thus R_θ is measure-preserving.

One good way of thinking about this transformation R_θ is by thinking of X as a circle: take the half-open interval $[0,1)$, and connect the missing endpoint 1 to the present endpoint 0, so that it wraps around. Then R_θ is a rotation by θ on the circle, where θ is the proportion of the circumference of the circle. This is often a good way of parametrizing the circle, so that the full circle has measure 1.

In the future, we'll write $\mathbb{R}/\mathbb{Z}$ for X: it consists of the real numbers, except that two real numbers are considered equal if they differ by an integer.

Example 4.4. Let $X = \{0,1\}^{\mathbb{N}}$ be the set of infinite binary sequences, i.e. sequences $(a_0, a_1, a_2, \ldots)$ where each $a_i \in \{0,1\}$. The σ-algebra on X is the smallest one containing all *cylinder sets*. For a nonnegative integer n and $b_0, \ldots, b_n \in \{0,1\}$ the corresponding cylinder set is $S_{b_0,\ldots,b_n} = \{(a_0, a_1, \ldots) : a_0 = b_0, \ldots, a_n = b_n\}$. So this means that we constrain the values of the first several elements of the sequence. We define the measure of a cylinder set $S_{b_0,\ldots,b_n}$ to be $\frac{1}{2^{n+1}}$. (The $n+1$ is because there are $n+1$ of the b_i's, rather than just n.)

A typical example of a measure-preserving transformation on X is the *left-shift transformation*, also known as the *Bernoulli shift*. This means that $T(a_0, a_1, a_2, \ldots) = (a_1, a_2, a_3, \ldots)$, so that we throw away the first (or zeroth) term and shift everything else one position to the left. In order to see that T is measure-preserving, note first that if $S_{b_0,b_1,\ldots,b_n}$ is any cylinder set, then $T^{-1}(S_{b_0,b_1,\ldots,b_n}) = S_{0,b_0,b_1,\ldots,b_n} \cup S_{1,b_0,b_1,\ldots,b_n}$ is a union of two cylinder sets and hence measurable. Now we have to check that it preserves the measure. If we have any cylinder set $S_{b_0,\ldots,b_n}$, then $\mu(S_{b_0,\ldots,b_n}) = \frac{1}{2^{n+1}}$, whereas $T^{-1}(S_{b_0,\ldots,b_n})$ is a union of two disjoint cylinder sets, each of measure $\frac{1}{2^{n+2}}$, so its total measure is again $\frac{1}{2^{n+1}}$. Thus T is indeed measure-preserving, at least on cylinder sets. As we will see soon, this is enough: if T preserves the measure of all cylinder sets, then it preserves the measure of all measurable sets; see Problem 2. We'll discuss this in more detail in a different example soon.

Example 4.5. We can also consider more general types of Bernoulli shifts. Let $(Y, \mathcal{B}, \nu)$ be any probability space, and define a new space $X = Y^{\mathbb{N}}$ to be the set of sequences of elements of Y. Define the *cylinder sets* on X as follows: let $B_0, \ldots, B_n$ be any $\mathcal{B}$-measurable sets, and define the cylinder set

$$S_{B_0,\ldots,B_n} = \{(y_0, y_1, \ldots) : y_i \in B_i \text{ for } 0 \leq i \leq n\}.$$

We now put a σ-algebra $\mathcal{A}$ on X, which is the smallest σ-algebra containing all the cylinder sets. The measure μ of a cylinder set is defined to be

$$\mu(S_{B_0,\ldots,B_n}) = \prod_{i=0}^{n} \nu(B_i).$$

We once again consider the transformation T on $(X, \mathcal{A}, \mu)$ to be the Bernoulli shift

$$T(y_0, y_1, \ldots) = (y_1, y_2, \ldots).$$

This transformation preserves the measure of all cylinder sets, because we have

$$T^{-1}(S_{B_0,\ldots,B_n}) = S_{Y,B_0,\ldots,B_n},$$

and these two sets have the same measure since $\nu(Y) = 1$. As before, this implies that T is measure-preserving.

Example 4.6. Let $X = [0,1] \times [0,1]$ be a square, and let $\mathcal{A}$ and μ be the Lebesgue σ-algebra and Lebesgue measure on X. The *baker's transformation* on X is the function $T : X \to X$ defined by

$$T(x, y) = \begin{cases} (2x, \frac{y}{2}) & 0 \leq x < \frac{1}{2}, \\ (2x - 1, \frac{y+1}{2}) & \frac{1}{2} \leq x \leq 1. \end{cases}$$

See Figure 4.1 for a pictorial interpretation of the baker's transformation. It is so called because it is reminiscent of what a baker does to mix dough, although that one is a bit more complicated because of the presence of a third dimension. There are a number of things to check in order to prove rigorously that the baker's transformation is measure-preserving, but the basic idea is to check it on a rectangle. If R is a rectangle in either the dark

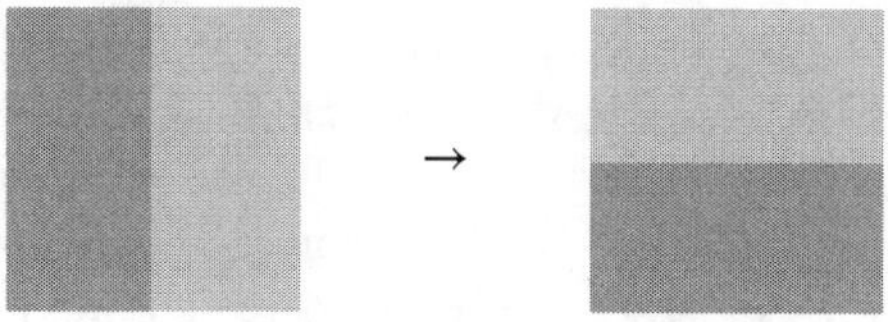

Figure 4.1. The baker's transformation

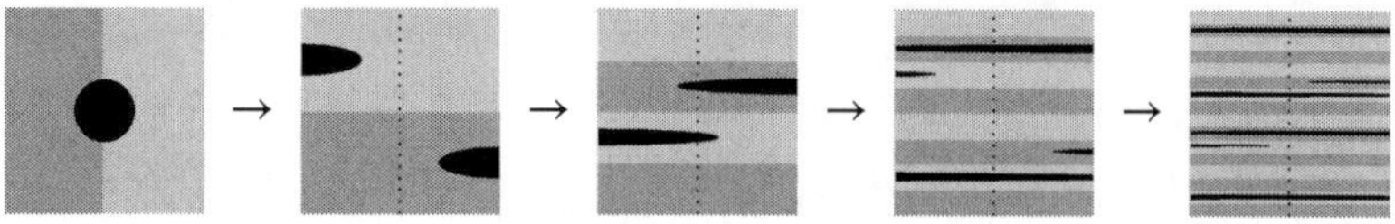

Figure 4.2. Iterations of the baker's transformation

or light region, then $T^{-1}(R)$ is a rectangle that's half as wide and twice as tall, and thus has the same area. In Problem 3, you will work out the details to show that the baker's transformation is measure-preserving.

Why is this transformation relevant to bakers? This is due to the fact that iteration of this transformation leads to *mixing*, a topic we will study in Chapter 8. Imagine that you want to make a 2-dimensional loaf of bread, in the shape of a square. So, you make the dough and form it into a square. But now you want to add something to the bread, perhaps some raisins. One way you can do this is to put all the raisins somewhere, then apply the baker's transformation many times. You can see a picture of this process in Figure 4.2. It may take a while, but eventually the raisins get spread out evenly throughout the entire dough. It's also worth noting that it would probably be more convenient for a baker to do a fold rather than a cut, so that the top (blue) rectangle ends up upside-down. This is true based on practical considerations, but mathematically, it doesn't make much of a difference.

You might wonder why we require that $\mu(A) = \mu(T^{-1}(A))$ rather than $\mu(A) = \mu(T(A))$. In the latter case, the Bernoulli shift wouldn't be a measure-preserving transformation. To see this, let $A = S_0 = \{(a_0, a_1, \ldots) : a_0 = 0\}$ be the cylinder set. This is a measurable set, and $T(A) = \{0,1\}^{\mathbb{N}}$. Thus we have $\mu(A) = \frac{1}{2}$ and $\mu(T(A)) = 1$. We'd still like to consider T to be a measure-preserving transformation, though, and admit that A wasn't quite the right sort of set to be considering, since it's not the preimage of any set.

However, many of the transformations we consider will be invertible, i.e. it is a bijection: for every $y \in X$, there is a unique $x \in X$ such that $T(x) = y$. For invertible transformations, the condition $\mu(A) = \mu(T^{-1}(A))$ for all A is equivalent to the condition $\mu(A) = \mu(T(A))$ for all A (see Problem 9), so we're free to use that one instead. Of the transformations we've seen so far, the rotation and the baker's transformation are invertible, whereas the Bernoulli shift is not.

4.2 Digital Expansions

One of the big goals of this book will be to investigate the digital representations of "typical" real numbers. Consider a real number $\alpha \in [0, 1)$. We can look at its decimal expansion and ask about its distribution of digits. Does it have the same number of 0's, 1's, 2's, and so forth? What does that even mean?

Let's fix some notation and write $\alpha = .a_1 a_2 a_3 \cdots$, where each a_i satisfies $0 \le a_i \le 9$. What it ought to mean for the decimal expansion of α to have the same number of each digit is that, if we look at the first N digits of the decimal expansion of α, there should be close to the same number of 0's, 1's, 2's, and so forth. More precisely, let us define a function $\theta(d, N, \alpha)$ to be

$$\theta(d, N, \alpha) = \#\{i : 1 \le i \le N : a_i = d\},$$

where the a_i's are the digits of α. The precise statement for what it means for each digit to occur equally often is that

$$\lim_{N\to\infty} \frac{\theta(d, N, \alpha)}{N} = \frac{1}{10}$$

for each $0 \le d \le 9$. In particular, that requires the limit to exist.

Definition 4.7. Let $\alpha \in [0, 1)$. We say that α is *simply normal to base 10* if, in its decimal expansion, we have

$$\lim_{N\to\infty} \frac{\theta(d, N, \alpha)}{N} = \frac{1}{10}$$

for all d with $0 \le d \le 9$.

More generally, for any base $b \geq 2$, we have the notion of a number which is simply normal to base b: just replace the decimal expansion with base-b expansion and the $\frac{1}{10}$ with $\frac{1}{b}$ (and the $0 \leq d \leq 9$ with $0 \leq d \leq b-1$).

Once we have developed some of the tools of ergodic theory, we'll be able to prove the following theorem:

Theorem 4.8. *Let* $A_b = \{\alpha \in [0,1) : \alpha \text{ is simply normal to base } b\}$. *Then* $\lambda(A_b) = 1$, *where* λ *denotes the Lebesgue measure on* $\mathbb{R}/\mathbb{Z}$.

In fact, we'll be able to prove something even stronger. Not only does every digit occur equally often, but so does every window of length 2, every window of length 3, and so forth, in a set of measure 1. That is, not only should there be roughly the same number of 5's and 8's, but there should also be roughly the same number of consecutive 28's as 55's, roughly the same number of 010's as 691's, and so forth. A number satisfying that property is said to be *normal to base 10*, with the obvious extension to normality to base b.

As we shall see, our tools will be good enough to prove that most numbers are normal to all bases. However, we're out of luck if we're interested in a particular number. If we want to know whether π (or, more precisely, $\pi - 3$) is normal, we have no idea how to do that. Determining the normality of famous numbers like π is a celebrated open problem. There are some numbers that are known to be normal (say, to base 10), but only because they are constructed explicitly for this purpose. The basic example is *Champernowne's constant*

$$0.12345678910111213141516\cdots.$$

It's built to be normal in base 10 (see [**Cha33**]), but it's an open problem whether it is normal in other bases. One can show that it's normal in bases that are powers of 10, but other bases are still completely open.

So, what does all this have to do with measure-preserving transformations? We will analyze the normality of typical numbers by means of the measure-preserving transformation $T : \mathbb{R}/\mathbb{Z} \to \mathbb{R}/\mathbb{Z}$ defined by $T(x) = 10x - \lfloor 10x \rfloor$. This should remind you of the Bernoulli shift: just chop off the first digit in the decimal expansion of x and shift everything else over one place to the right. For instance, $T(\pi - 3) = .4159265358979\cdots$; we just removed the initial 1 from the decimal expansion of $\pi - 3$.

Let's see why T is measure-preserving. It may seem a bit daunting to check, because we have to show that for *every* Lebesgue-measurable subset $A \subseteq \mathbb{R}/\mathbb{Z}$, we have $\lambda(T^{-1}(A)) = \lambda(A)$. But in fact, we don't really have to check all of them, we only have to check it on subsets that approximate all Lebesgue-measurable subsets sufficiently well. In particular, consider the collection $\mathcal{C}$ of intervals of the form $\left[\frac{a}{10^n}, \frac{a+1}{10^n}\right)$, where $0 \leq a < 10^n$. Then any $A \in \mathcal{L}$ can be approximated arbitrarily well by a countable union of elements of $\mathcal{C}$, in the sense that for all $A \in \mathcal{L}$,

$$\lambda(A) = \inf\left\{\sum_{j=1}^{\infty} \lambda(I_j) \,:\, A \subseteq \bigcup_{j=1}^{\infty} I_j \text{ and } I_j \in \mathcal{C} \text{ for all } j\right\}.$$

The set $\mathcal{C}$ is called a *sufficient semiring* for the Lebesgue σ-algebra $\mathcal{L}$.

Why does it suffice to check that T is measure-preserving on our sufficient semiring $\mathcal{C}$? Take any $A \in \mathcal{L}$. Then we can find a collection of measurable subsets $H_1, H_2, \ldots$, each of which is a countable union of disjoint elements of $\mathcal{C}$, such that the H_i's are decreasing in the sense that $H_1 \supseteq H_2 \supseteq \cdots$, and each $H_i \supseteq A$, and if $H = \bigcap_{j=1}^{\infty} H_j$, then $\lambda(H \setminus A) = 0$. (This is actually true for all sufficient semirings, as you will check in Problem 2.)

Let's show that, under this setup, $\lambda(T^{-1}(H)) = \lambda(H)$. Let's write each H_n as

$$H_n = \bigsqcup_{i=1}^{\infty} C_{n,i},$$

with each $C_{n,i} \in \mathcal{C}$. Then $T^{-1}(H_n)$ is measurable, because $T^{-1}(C_{n,i})$ is a union of 10 disjoint intervals, so $T^{-1}(H_n)$ is also a disjoint union of countably many intervals. Furthermore, on each $C_{n,i}$, $T^{-1}(C_{n,i})$ is a disjoint union of 10 intervals, each of which has length $\left(\frac{1}{10}\right)^{\text{th}}$ that of $C_{n,i}$, so T is measure-preserving on $C_{n,i}$ and hence on H_n.

Next, we show that T is measure-preserving on H: we have

$$\begin{aligned}\lambda(T^{-1}(H)) &= \lambda\left(\bigcap_{n=1}^{\infty} T^{-1}(H_n)\right) \\ &= \lim_{n\to\infty} \lambda(T^{-1}(H_n)) \\ &= \lim_{n\to\infty} \lambda(H_n) \\ &= \lambda\left(\bigcap_{n=1}^{\infty} H_n\right) \\ &= \lambda(H).\end{aligned}$$

Now, let $N = H \setminus A$, so that N is a null set, i.e. $\lambda(N) = 0$. We can play the same game as before, finding an H' serving the same role for N as H served for A, so that in particular $\lambda(H' \setminus N) = 0$, and by the argument from before, T is measure-preserving on H'. But if $\lambda(H' \setminus N) = 0$ and $\lambda(N) = 0$, then $\lambda(H') = 0$. Thus $\lambda(T^{-1}(H')) = 0$, and since $N \subseteq H'$, $\lambda(T^{-1}(N)) = 0$.

Finally, we have $T^{-1}(A) = T^{-1}(H) \setminus T^{-1}(N)$, so

$$\lambda(T^{-1}(A)) = \lambda(T^{-1}(H)) - \lambda(T^{-1}(N)) = \lambda(T^{-1}(H)) = \lambda(H) = \lambda(A),$$

as desired.

This is a typical technique used to show that a transformation is measure-preserving: just check that it is measure-preserving on a sufficient semiring, which is much easier because the elements of the sufficient semiring, and ideally also their preimages, can be concretely described.

4.3 Continued Fractions

One way of representing real numbers is using their decimal (or base-b) expansion. Arguably a better way is using their continued fraction representations. A *continued fraction* is an expression of the form

$$a_0 + \cfrac{1}{a_1 + \cfrac{1}{a_2 + \cfrac{1}{a_3 + \cfrac{1}{a_4 + \cdots}}}}, \tag{4.1}$$

where $a_0 \in \mathbb{Z}$, and all the other a_i's are positive integers. Such an expression might end at some point, such as

$$3 + \cfrac{1}{6 + \cfrac{1}{1 + \cfrac{1}{3}}},$$

or it might go on forever. For typographical reasons, we'll write

$$[a_0; a_1, a_2, a_3, a_4, \ldots]$$

in place of (4.1).

In order to evaluate a finite continued fraction, the easiest thing to do is to start from the bottom and work your way up. So, for instance, we have

$$3 + \cfrac{1}{6 + \cfrac{1}{1 + \cfrac{1}{3}}} = 3 + \cfrac{1}{6 + \cfrac{1}{4/3}} = 3 + \cfrac{1}{6 + \frac{3}{4}} = 3 + \frac{1}{27/4} = 3 + \frac{4}{27} = \frac{85}{27},$$

and we're done. To evaluate an infinite continued fraction, cut it off at further and further points, evaluate the finite ones, and take the limit.

The theory of continued fractions is fascinating and is deeply connected to many topics in number theory. See for instance [**Khi64**] for a discussion of the use of continued fractions in Diophantine approximation as well as measure-theoretic aspects of them. Suffice it to say for now that rational numbers have finite continued fraction representations, quadratic irrationals (i.e. real roots of irreducible quadratic polynomials with integer coefficients, like $\frac{3+8\sqrt{7}}{19}$), have *eventually periodic* continued fraction representations, i.e. there is some positive integer

p so that $a_{n+p} = a_n$ for all sufficiently large n. Most other numbers have random-looking continued fraction expansions, but some interesting numbers, like e, have predictable ones: we have

$$e = [2; 1, 2, 1, 1, 4, 1, 1, 6, 1, 1, 8, 1, 1, 10, \ldots].$$

See [**Coh06**] for a proof.

For any real number α, we can construct a continued fraction equal to α. Let's see how to do this with an example, say of π. To do it, we let $a_0 = \lfloor \pi \rfloor = 3$. The continued fraction representation of π will be $[3; \ldots]$. Next, we subtract 3 from π and take the reciprocal. We then repeat the process. Thus $a_1 = \left\lfloor \frac{1}{\pi - 3} \right\rfloor$. Continuing on this way, we find that the continued fraction expansion of π is $[3; 7, 15, 1, 292, 1, 1, 1, 2, 1, 3, 1, 14, 2, 1, 1, 2, 2, \ldots]$. If we cut it off after each a_i, we get the following sequence of rational numbers:

$$3, \frac{22}{7}, \frac{333}{106}, \frac{355}{113}, \frac{103993}{33102}, \ldots$$

These are the best approximations of π with at most those denominators, which is why some of the famous approximations for π appear on this list.

The transformation here is as follows: our space is again $[0, 1)$. The transformation T is $T : [0, 1) \to [0, 1)$ defined by $T(x) = \frac{1}{x} - \left\lfloor \frac{1}{x} \right\rfloor$, provided that $x \neq 0$, and $T(0) = 0$. Given a number $x \in [0, 1)$, let us write $x_0 = x$ and $x_{n+1} = T(x_n)$ for $n \geq 0$. Furthermore, let's write $a_{n+1} = \left\lfloor \frac{1}{x_n} \right\rfloor$. This is relevant for continued fractions because we have

$$x_0 = \frac{1}{a_1 + x_1} = \cfrac{1}{a_1 + \cfrac{1}{a_2 + x_2}} = \cfrac{1}{a_1 + \cfrac{1}{a_2 + \cfrac{1}{a_3 + x_3}}} = \cdots,$$

at least when $x \notin \mathbb{Q}$ (so as to prevent us from ever reaching 0). In particular, assuming that $x \notin \mathbb{Q}$, the continued fraction expansion for x is $[0; a_1, a_2, a_3, a_4, \ldots]$.

Is this transformation measure-preserving? Let's check a special case. For $x \in [0, 1)$, we have $a_1 = 1$ if and only if $x \in \left(\frac{1}{2}, 1\right)$. So, it's logical to look at $T^{-1}\left(\left(\frac{1}{2}, 1\right)\right)$. When is $x \in T^{-1}\left(\left(\frac{1}{2}, 1\right)\right)$? This means that $T(x) \in \left(\frac{1}{2}, 1\right)$, i.e. $\frac{1}{x} - \left\lfloor \frac{1}{x} \right\rfloor \in \left(\frac{1}{2}, 1\right)$. There are infinitely ways this can happen, depending on what $\left\lfloor \frac{1}{x} \right\rfloor = a_1$ is. If $a_1 = k$, then that means that $x \in \left(\frac{1}{k+1}, \frac{1}{k}\right]$. If in addition $T(x) \in \left(\frac{1}{2}, 1\right)$, that means that

$$1 \leq \frac{1}{1/x - k} < 2,$$

or

$$\frac{1}{k+1} \leq x < \frac{1}{k+1/2}.$$

The Lebesgue measure of that interval is $\frac{1}{k+1/2} - \frac{1}{k+1}$. Summing over k from 1 to ∞, we find that

$$\begin{aligned} \lambda\left(T^{-1}\left(\left(\frac{1}{2}, 1\right)\right)\right) &= \sum_{k=1}^{\infty} \left(\frac{1}{k+1/2} - \frac{1}{k+1}\right) \\ &= 2\sum_{k=1}^{\infty} \left(\frac{1}{2k+1} - \frac{1}{2k+2}\right) \\ &= 2\log 2 - 1. \end{aligned}$$

Since that's not the same as $\lambda\left(\left(\frac{1}{2}, 1\right)\right) = \frac{1}{2}$, we conclude that T is *not* measure-preserving.

However, this is not the end of the story, because we don't have to use Lebesgue measure. Perhaps there's a different measure on $[0, 1)$ for which T is measure-preserving. Indeed, there is: this measure is known as Gauß measure. It is defined by

$$\mu(A) = \frac{1}{\log 2} \int_A \frac{1}{1+x}\, d\lambda,$$

where A is any Lebesgue-measurable subset of $[0, 1)$. Gauß didn't give any explanation for how he came up with this measure, but we can at least check that it is measure-preserving. By a similar analysis to before, if (a, b) is any open interval in $[0, 1)$, then we have

$$T^{-1}((a, b)) = \bigcup_{n=1}^{\infty} \left(\frac{1}{n+b}, \frac{1}{n+a}\right).$$

Thus we have

$$\mu((a, b)) = \frac{1}{\log 2} \int_a^b \frac{dx}{1+x} = \frac{1}{\log 2}(\log(1+b) - \log(1+a)),$$

whereas

$$
\begin{aligned}
\mu(T^{-1}((a,b))) &= \sum_{n=1}^{\infty} \frac{1}{\log 2} \int_{1/(n+b)}^{1/(n+a)} \frac{dx}{1+x} \\
&= \frac{1}{\log 2} \sum_{n=1}^{\infty} \left[\log\left(1 + \frac{1}{n+a}\right) - \log\left(1 + \frac{1}{n+b}\right) \right] \\
&= \frac{1}{\log 2} \log \prod_{n=1}^{\infty} \frac{(n+a+1)(n+b)}{(n+a)(n+b+1)} \\
&= \frac{1}{\log 2} \log \frac{b+1}{a+1} \\
&= \frac{1}{\log 2} (\log(1+b) - \log(1+a)).
\end{aligned}
$$

The two are equal, so T preserves the Gauß measures of intervals and hence all measurable sets by the sufficient semiring machinery discussed earlier.

Recall that we stated (and will later prove) that almost all numbers are normal, in that the set of non-normal numbers is a null set. There is an analogous result for continued fractions.

Theorem 4.9 (Khinchin). *Suppose the continued fraction for $\alpha \in [0,1)$ is $[0; a_1, a_2, a_3, \ldots]$. Let*

$$G_n(\alpha) = (a_1 a_2 \cdots a_n)^{1/n},$$

and let

$$G(\alpha) = \lim_{n\to\infty} G_n(\alpha),$$

if this limit exists. There is a constant K such that, for almost all $\alpha \in [0,1)$, $G(\alpha)$ exists and is equal to K.

The remarkable thing about Khinchin's Theorem is that this constant K, called *Khinchin's constant*, is independent of α. In fact, we can describe it explicitly:

$$K = \prod_{n=1}^{\infty} \left(1 + \frac{1}{n(n+2)}\right)^{\frac{\log n}{\log 2}} \approx 2.685452001.$$

We'll prove Khinchin's Theorem in Chapter 7, when we discuss invariant measures.

4.4 Lüroth Series

Let's analyze both the decimal expansion and continued fraction representation of real numbers in $[0, 1)$ with the goal of finding other similar representations. In decimal expansions, we break the interval $[0, 1)$ into 10 pieces, namely the intervals $I_k = \left[\frac{k}{10}, \frac{k+1}{10}\right)$ for $0 \leq k \leq 9$. Then we scale each of these intervals back so that it fills up the entire interval $[0, 1)$; that is, each of these intervals is mapped bijectively onto $[0, 1)$. That's what the transformation T given by $T(x) = 10x - \lfloor 10x \rfloor$ that we discussed in §4.2 does: the image of each I_k is all of $[0, 1)$. The decimal expansion of x records, for each n, which I_k the number $T^n(x)$ is in.

Continued fractions do something similar, but with a different way of breaking up $[0, 1)$. Here, we break up the interval $[0, 1)$ into subintervals $J_k = \left(\frac{1}{k+1}, \frac{1}{k}\right]$ (except that $J_1 = \left(\frac{1}{2}, 1\right)$ and $J_0 = \{0\}$). Each one of these intervals, except for J_0, gets mapped back to the entire interval $[0, 1)$ by means of the continued fraction transformation $T(x) = \frac{1}{x} - \left\lfloor \frac{1}{x} \right\rfloor$, and this mapping is a bijection. (Again, there's a small exception for J_1, where the mapping isn't surjective because its image does not include 0.) Here the mapping isn't a pure scaling like it is in the decimal expansion, but it's a more complicated mapping that involves stretching differently at different points. But that's okay, it's still a bijective mapping from each of the J_k's to $[0, 1)$.

Another less well-known way of representing numbers in $[0, 1)$ is using their *Lüroth series expansions*. To do this, we break up $[0, 1)$ almost as we do in continued fractions, with the same intervals J_k (except here we'll use the other half-open intervals $J'_k = \left[\frac{1}{k+1}, \frac{1}{k}\right)$), but this time we apply at each stage a transformation that scales each interval linearly to take up the entire interval $[0, 1)$. As a formula, this is the transformation

$$T(x) = \begin{cases} k(k+1)x - k & k \in J'_k = \left[\frac{1}{k+1}, \frac{1}{k}\right), \\ 0 & x = 0, \end{cases} \tag{4.2}$$

or alternatively

$$T(x) = \left\lfloor \frac{1}{x} \right\rfloor \left\lfloor \frac{1}{x} + 1 \right\rfloor x - \left\lfloor \frac{1}{x} \right\rfloor.$$

So, each interval gets stretched uniformly. We then get an expansion out of T by recording which J'_k we're in at each iteration. We'll actually shift it by 1, so we'll let $a_n = k + 1$ if $T^{n-1}(x) \in J'_k$.

Another way of saying this is that we can represent an arbitrary $x \in [0,1)$ as

$$x = \frac{1}{a_1} + \frac{1}{a_1(a_1-1)a_2} + \frac{1}{a_1(a_1-1)a_2(a_2-1)a_3} + \frac{1}{a_1(a_1-1)a_2(a_2-1)a_3(a_3-1)a_4} + \cdots,$$

where the sum can be finite or infinite. Here each a_k is an integer ≥ 2.

Example 4.10. Let us compute the first few terms of the Lüroth series for $\pi - 3$. Since $\pi - 3 \in \left[\frac{1}{8}, \frac{1}{7}\right)$, we have $a_1 = 8$. To do the next iteration, we have to apply T to $\pi - 3$, to get $56\pi - 175 \approx .9291886$. Since $T(\pi-3) \in \left[\frac{1}{2}, 1\right)$, we have $a_2 = 2$. Continuing on in this way, we get $a_3 = 2$, $a_4 = 2$, $a_5 = 3$, $a_6 = 2$, $a_7 = 5$, $a_8 = 24$, and so forth. In terms of a series, this means that

$$\pi - 3 = \frac{1}{8} + \frac{1}{8 \times 7 \times 2} + \frac{1}{8 \times 7 \times 2 \times 1 \times 2} + \frac{1}{8 \times 7 \times 2 \times 1 \times 2 \times 1 \times 2} + \cdots.$$

The first few partial sums are

.125, .1339, .1384, .1406, .141369, .141555, .141592262.

These aren't as impressive as the convergents of the continued fraction, by any means, but they're still converging to $\pi - 3$, as we can see.

Like the decimal expansion transformation $T(x) = 10x - \lfloor 10x \rfloor$, but unlike the continued fraction transformation $T(x) = \frac{1}{x} - \left\lfloor \frac{1}{x} \right\rfloor$, the Lüroth transformation of (4.2) is measure-preserving with respect to the Lebesgue measure, as you will show in Problem 11.

4.5 Problems

(1) Prove that if T is measure-preserving on $(X, \mathcal{A}, \mu)$, then T^n is also measure-preserving for each integer $n \geq 2$.

(2) Let $(X, \mathcal{A}, \mu)$ be a measure space. A subset $\mathcal{C}$ of $\mathcal{A}$ is said to be a *sufficient semiring* if it satisfies the following properties:

- $\mathcal{C}$ is nonempty.

- If $A, B \in \mathcal{C}$, then $A \cap B \in \mathcal{C}$.
- If $A, B \in \mathcal{C}$, then there are finitely many disjoint sets $C_1, \ldots, C_n \in \mathcal{C}$ such that $A \setminus B = C_1 \sqcup C_2 \sqcup \cdots \sqcup C_n$.
- For every $A \in \mathcal{C}$, $\mu(A) < \infty$.
- For every $A \in \mathcal{A}$, we have

$$\mu(A) = \inf\left\{\sum_{j=1}^{\infty} \mu(C_j) : A \subseteq \bigcup_{j=1}^{\infty} C_j \text{ and } C_j \in \mathcal{C} \text{ for all } j\right\}.$$

Prove that if T is a transformation on $(X, \mathcal{A}, \mu)$ and $\mathcal{C}$ is a sufficient semiring, and $\mu(T^{-1}(C)) = \mu(C)$ for all $C \in \mathcal{C}$, then T is measure-preserving.

(3) Prove that the baker's transformation is measure-preserving.

(4) Find an example of a transformation T on a measure space $(X, \mathcal{A}, \mu)$ such that T is not measure-preserving, but T^n is for some integer $n \geq 2$. Can you find such an example for each integer $n \geq 2$? (That is, find a different example for each n.)

(5) Show that if $(X, \mathcal{A}, \mu, T)$ is a measure-preserving system, and $f \in L^1(\mu)$, then

$$\lim_{n\to\infty} \frac{f(T^n(x))}{n} = 0$$

for almost all $x \in X$.

(6) Let T be the Lüroth series transformation of (4.2). For an interval $(a, b) \subset [0, 1)$, describe the preimage $T^{-1}((a, b))$.

(7) Let $\beta > 1$ be a real number which is not an integer. Prove that the transformation $T_\beta : [0, 1) \to [0, 1)$ defined by $T(x) = \beta x - \lfloor \beta x \rfloor$ is *not* measure-preserving with respect to the Lebesgue measure on $[0, 1)$.

(8) Even though the transformation T_β from Problem 7 is not measure-preserving with respect to the Lebesgue measure, it is still interesting, as it is related to expansions of numbers in base β. For a number $x \in [0, 1)$ and an integer $n \geq 1$, define $d_n(x) = \lfloor \beta T_\beta^{n-1}(x) \rfloor$. Prove that

$$x = \sum_{n=1}^{\infty} \frac{d_n(x)}{\beta^n}.$$

(9) Let T be a bijective transformation on a measure space $(X, \mathcal{A}, \mu)$. Prove that T is measure-preserving if and only if $\mu(T(A)) = \mu(A)$ for all $A \in \mathcal{A}$. (Note that the definition of measure-preserving says that $\mu(T^{-1}(A)) = \mu(A)$.)

(10) Prove that the Boole transformation $T : \mathbb{R} \to \mathbb{R}$ defined by $T(x) = x - \frac{1}{x}$ for $x \neq 0$ and $T(0) = 0$ is measure-preserving with respect to the Lebesgue measure.

(11) Prove that the Lüroth transformation from (4.2) is a measure-preserving transformation on $(\mathbb{R}/\mathbb{Z}, \mathcal{L}, \lambda)$.

(12) Prove that a rational number $x \in [0, 1)$ has a finite or periodic Lüroth series expansion.

(13) Let $(X, \mathcal{A}, \mu)$ be a measure space, and let T be a transformation on it. Prove that T is measure-preserving if and only if

$$\int_X f \, d\mu = \int_X f \circ T \, d\mu$$

for all $f \in L^1(\mu)$.

(14) Find a transformation T on $(\mathbb{R}/\mathbb{Z}, \mathcal{L})$ for which there are two probability measures μ and ν such that T is measure-preserving with respect to both μ and ν.

(15) How do you generalize the baker's transformation to $[0, 1]^3$ such that the iterates are well-mixed? (We will define the notion of mixing formally in Chapter 8.)

Chapter 5

The Poincaré Recurrence Theorem

5.1 Recurrence

Suppose $(X, \mathcal{A}, \mu, T)$ is a measure-preserving system. The sort of thing we're usually interested in in ergodic theory is studying the behavior of the iterates of T, i.e. the sequence $x, T(x), T^2(x), T^3(x), \ldots$ A logical question to ask is whether some iterate of x is equal to x again.

As an example, consider the rotation R_α on $\mathbb{R}/\mathbb{Z}$. The nth iterate of R_α applied to x is $R_\alpha^n(x) = x + n\alpha \pmod 1$. In order for this to be equal to x, we must have $x + n\alpha \equiv x \pmod 1$, i.e. $n\alpha \equiv 0 \pmod 1$. If α is rational, then this does indeed happen. Suppose $\alpha = \frac{p}{q}$. Then if $n = q$, we have $n\alpha \equiv 0 \pmod 1$, so the nth iterate of R_α is just the identity map, not only for x but for all of $\mathbb{R}/\mathbb{Z}$.

On the other hand, if α is irrational, then $n\alpha \not\equiv 0 \pmod 1$ for all positive integers n, so no iterate of R_α applied to x is equal to x. But then, maybe we don't care so much that some $R_\alpha^n(x)$ is *exactly* equal to x. Perhaps we'd only like to know whether some $R_\alpha^n(x)$ is *close* to x, where you can decide how close you want to require. For example, is there some $R_\alpha^n(x)$ that is within 10^{-10} of x?

Measure theory gives us exactly the right tools for answering such questions. The difference between getting back to x on the nose and getting back to a small interval around x is that the former is a null set, whereas the latter has positive measure. If we have a set of positive measure, we can hope to return to it, whereas if we have a null set, the iterates can just dance around, barely missing the starting set every time.

Now, sometimes it happens that iterates of a point really can miss a set of positive measure. Consider the continued fraction transformation on $[0, 1)$ that we discussed in Chapter 4. Recall that it is defined by

$$T(x) = \begin{cases} \frac{1}{x} - \left\lfloor \frac{1}{x} \right\rfloor & x \neq 0, \\ 0 & x = 0. \end{cases}$$

Suppose we start at $x = \frac{1}{2}$. Do we ever get back close to $\frac{1}{2}$? Let's check the iterates of $\frac{1}{2}$. We have $T(\frac{1}{2}) = 0$, and then $T(0) = 0$, so all iterates of $\frac{1}{2}$ are 0. Thus we never return close to our starting point.

This isn't just true for rational numbers. For an example with an irrational number, start with $x = \frac{1+\sqrt{2}}{5}$. Here are the iterates of x:

$$T(x) = 5\sqrt{2} - 7, \qquad T^2(x) = 5\sqrt{2} - 7, \qquad \dots$$

So, all iterates are the same, and they never come back to the starting point, or anywhere close.

But this turns out to be rare behavior, as we shall see. For *most* starting points, we eventually get back near the starting point.

Let's now state the precise definition of the notion we have been describing, namely recurrence.

Definition 5.1. A measure-preserving transformation T on a measure space $(X, \mathcal{A}, \mu)$ is said to be *recurrent* if for any $A \in \mathcal{A}$ with $\mu(A) > 0$, there is a subset $N \in \mathcal{A}$ with $N \subseteq A$ and $\mu(N) = 0$ such that for all $x \in A \setminus N$, there is some positive integer n (depending on x) such that $T^n(x) \in A$.

That's a bit of a mouthful, so let's take a moment to parse what it means. First, suppose that N is the empty set. Then this is just saying that for all $x \in A$, some iterate $T^n(x)$ of x lies in A. But this is too strict of a criterion, because as we saw with continued fractions, it doesn't hold. So instead, we don't require that *every* $x \in A$ has an iterate lying in A. We only require that this happens for *almost all* $x \in A$, i.e. all but a null set.

We think about n as being the amount of time it takes for x to return back to A. There's no reason that different points in A have to take the same amount of time to return to A, which is why we allow n to depend on x.

Not all measure-preserving transformations are recurrent. A typical example of a non-recurrent measure-preserving transformation is $T : \mathbb{R} \to \mathbb{R}$ defined by $T(x) = x + 1$. If we take A to be a small interval, say $\left(0, \frac{1}{2}\right)$, then no iterate of any point in the interval will ever lie in the interval. Thus this transformation is not recurrent. Of course, this measure space isn't finite, and we shall see that that is what makes all the difference.

Note that the definition of recurrence only says that x must return to A once. We might instead want x to return to A infinitely many times. It seems obvious that if it returns to A once then it returns to A infinitely many times, as follows: let $x_1 = T^n(x)$, where n is chosen so that $T^n(x) \in A$. Then let $x_2 = T^{n_1}(x_1)$, where n_1 is chosen so that $T^{n_1}(x_1) \in A$. Then let $x_3 = T^{n_2}(x_2)$, where n_2 is chosen so that $T^{n_2}(x_2) \in A$. The problem is that we have these null sets to worry about, so how do we know that we won't eventually end up in the measure-zero set, only it takes a different amount of time to get there for different starting points? It turns out that we needn't be worried about such a thing, but this requires a proof.

Lemma 5.2. *Let T be a recurrent measure-preserving transformation on $(X, \mathcal{A}, \mu)$. Then for all $A \in \mathcal{A}$ with $\mu(A) > 0$, there is a null set $N \subseteq A$ such that for all $x \in A \setminus N$, there is an infinite increasing sequence $n_1 < n_2 < n_3 < \cdots$ such that $T^{n_i}(x) \in A$ for all i.*

Proof. Since T is assumed to be recurrent, there is some null set N_1 such that for all $x \in A \setminus N_1$, there is some n depending on x such that $T^n(x) \in A$. Let

$$N = \bigcup_{k=0}^{\infty} T^{-k}(N_1).$$

Since T is measure-preserving, we have $\mu(T^{-k}(N_1)) = \mu(N_1) = 0$ for all k. Thus we have

$$\mu(N) = \mu\left(\bigcup_{k=0}^{\infty} T^{-k}(N_1)\right) \leq \sum_{k=0}^{\infty} \mu(T^{-k}(N_1)) = \sum_{k=0}^{\infty} 0 = 0,$$

so $\mu(N) = 0$. If $x \in A \setminus N \subseteq A \setminus N_1$, then by construction of N there is some n_1 depending on x such that $x_1 := T^{n_1}(x) \in A \setminus N$, since N already contains all points in X that have an iterate lying in N, or N_1. Since $x_1 \in A \setminus N$, we can repeat the argument, finding an n_2 such that $x_2 = T^{n_2}(x_1) \in A \setminus N$. We can continue this way, finding n_3, n_4, and so on. ■

Now, as we already mentioned, it can take different amounts of time for different points in A to return to A. But perhaps there's some amount of time that allows "a lot" of A to return to A. And indeed there is.

Definition 5.3. A measure-preserving transformation T on $(X, \mathcal{A}, \mu)$ is said to be *conservative* if for any set $A \in \mathcal{A}$ with $\mu(A) > 0$, there is some n such that $\mu(A \cap T^{-n}(A)) > 0$.

Lemma 5.4. *A measure-preserving transformation T on $(X, \mathcal{A}, \mu)$ is recurrent if and only if it is conservative.*

Proof. First, note that T is recurrent if and only if for all sets A,

$$\mu\left(A \setminus \bigcup_{n=1}^{\infty} T^{-n}(A)\right) = 0.$$

This is true because, if $\mu(A) = 0$, then certainly $\mu(A \setminus \bigcup_{n=1}^{\infty} T^{-n}(A)) = 0$ as well, the latter set being a subset of the former. But if $\mu(A) > 0$, then $A \setminus \bigcup_{n=1}^{\infty} T^{-n}(A)$ is the set of points in A that do not ever return to A after any positive number of iterations, which is a null set by the definition of recurrence.

Suppose that T is recurrent. We will prove that it is also conservative. Let A be a set with $\mu(A) > 0$. Then we have

$$\begin{aligned} 0 &= \mu\left(A \setminus \bigcup_{n=1}^{\infty} T^{-n}(A)\right) \\ &= \mu\left(A \setminus \bigcup_{n=1}^{\infty} [A \cap T^{-n}(A)]\right) \\ &\geq \mu(A) - \sum_{n=1}^{\infty} \mu(A \cap T^{-n}(A)). \end{aligned}$$

Thus there must be some $n > 0$ such that $\mu(A \cap T^{-n}(A)) > 0$. Thus T is conservative.

Now suppose that T is conservative, and let A be a set with $\mu(A) > 0$. Let

$$B = A \setminus \bigcup_{n=1}^{\infty} T^{-n}(A).$$

In other words, B is the set of points in A that never return to A. If $\mu(B) > 0$, then because T is conservative, there is some $n \geq 1$ such that $\mu(B \cap T^{-n}(B)) > 0$. In particular, there must be some point $x \in B \cap T^{-n}(B)$. But this contradicts the definition of B. Thus $\mu(B) = 0$, so T is recurrent. ■

5.2 The Poincaré Recurrence Theorem

It's now time to discuss the first serious theorem in ergodic theory and some of its applications.

Theorem 5.5 (Poincaré Recurrence Theorem). *Let $(X, \mathcal{A}, \mu, T)$ be a measure-preserving system. Then T is recurrent.*

Before proving the Poincaré Recurrence Theorem, let's remark on an extremely important special case. Suppose X is a finite set with $|X| = n$, $\mathcal{A} = \mathcal{P}(X)$ is the power set of X, and μ is the normalized counting measure, i.e. $\mu(A) = \frac{|A|}{n}$. In this case, a measure-preserving transformation is simply a bijective function. Let's now take a single element $x \in X$. Since $\mu(\{x\}) = \frac{1}{n} > 0$, Poincaré Recurrence says that there is some $m > 0$ such that $T^{-m}(x) = x$. But since T is bijective, we can replace the $-m$ with m, i.e. $T^m(x) = x$ (the conditions $T^m(x) = x$ and $T^{-m}(x) = x$ are equivalent when T is bijective). This is essentially the pigeonhole principle: consider the $n + 1$ elements

$$x, T(x), T^2(x), \ldots, T^n(x)$$

of X. Two of them, say $T^k(x)$ and $T^\ell(x)$ must be equal, say with $k < \ell$. But since T is a bijection, this means that $T^{-k}(T^k(x)) = T^{-k}(T^\ell(x))$, i.e. $x = T^{\ell-k}(x)$, so this proves Poincaré Recurrence in the case where X is a finite set. The proof in the general case is very much analogous to this argument.

Proof of the Poincaré Recurrence Theorem. It suffices to show that T is conservative. Let $A \in \mathcal{A}$ have $\mu(A) > 0$. Suppose that $\mu(A \cap T^{-n}(A)) = 0$ for all n. Let k, ℓ be positive integers with $\ell > k$, and let $n = \ell - k$. Then we have

$$\begin{aligned}
\mu(T^{-\ell}(A) \cap T^{-k}(A)) &= \mu(T^{-n-k}(A) \cap T^{-k}(A)) \\
&= \mu(T^{-k}(T^{-n}(A) \cap A)) \\
&= \mu(T^{-n}(A) \cap A) \\
&= 0
\end{aligned}$$

since T is measure-preserving. Thus the sets $T^{-n}(A)$ are almost pairwise disjoint, i.e. their pairwise intersections are all null sets. Thus

$$\mu\left(\bigcup_{n=0}^{\infty} T^{-n}(A)\right) = \sum_{n=0}^{\infty} \mu(T^{-n}(A)) = \sum_{n=0}^{\infty} \mu(A) = \infty.$$

But this is a contradiction, as $\mu(X)$ is assumed to be finite. Thus there must be some $n > 0$ such that $\mu(A \cap T^{-n}(A)) > 0$. ∎

We might be interested in knowing not just that $\mu(A \cap T^{-n}(A)) > 0$, but how large we can expect $\mu(A \cap T^{-n}(A))$ to be. For that we need a quantitative version of Poincaré recurrence. And here it is:

Theorem 5.6. *Let $(X, \mathcal{A}, \mu, T)$ be a measure-preserving system. Let $A \in \mathcal{A}$ have positive measure. Then for any $\varepsilon > 0$, there are infinitely many positive integers k such that $\mu(A \cap T^{-k}(A)) \geq \mu(A)^2 - \varepsilon$.*

This should seem pretty reasonable, in the context of a notion we'll be studying later called *mixing*. The idea is that after a while, the iterates of A, i.e. $T^n(A)$, should be spread out pretty evenly throughout X, so A and $T^n(A)$ (or A and $T^{-n}(A)$) should be independent of each other. If we think in terms of probability, we think of x as a random point, and x and $T^n(x)$ are eventually close to being independent random points of X. So, if $\mu(A)$ is the probability that a random point is in A, then $\mu(A \cap T^{-n}(A))$ should be roughly the probability that two independent random points are in A, which is $\mu(A)^2$. Of course, x and $T^n(x)$ aren't actually independent of each other, but the idea is that they should behave independently when n is large enough. This also isn't true without more hypotheses (that we will see later on), but here we only say that it's close enough to being true for infinitely many n.

Proof of Theorem 5.6. We apply the Cauchy–Schwarz Inequality, which as we recall says that if f and g are measurable functions, then

$$\left(\int_X fg\,d\mu\right)^2 \leq \left(\int_X f^2\,d\mu\right)\left(\int_X g^2\,d\mu\right).$$

For each positive integer N, let f_N be the function

$$f_N(x) = \sum_{n=1}^{N} \mathbb{1}_{T^{-n}(A)}(x),$$

i.e. $f_N(x) = |\{n : 1 \leq n \leq N, x \in T^{-n}(A)\}|$. We have

$$\int_X f_N\,d\mu = \sum_{n=1}^{N} \int_X \mathbb{1}_{T^{-n}(A)}\,d\mu = \sum_{n=1}^{N} \mu(T^{-n}(A)) = N\mu(A).$$

By the Cauchy–Schwarz Inequality, we have

$$\int_X f_N^2\,d\mu \geq \left(\int_X f_N\,d\mu\right)^2 = N^2\mu(A)^2.$$

We can rewrite the left side as

$$\begin{aligned}
\int_X f_N^2\,d\mu &= \int_X \left(\sum_{n=1}^{N} \mathbb{1}_{T^{-n}(A)}\right)^2 d\mu \\
&= \int_X \left(\sum_{n=1}^{N}\sum_{m=1}^{N} \mathbb{1}_{T^{-n}(A)\cap T^{-m}(A)}\right) d\mu \\
&= \sum_{n=1}^{N}\sum_{m=1}^{N} \mu(T^{-n}(A)\cap T^{-m}(A)).
\end{aligned}$$

Now, since T is measure-preserving, if $n \le m$, we have $\mu(T^{-n}(A) \cap T^{-m}(A)) = \mu(A \cap T^{-(m-n)}(A))$. Thus we have

$$\begin{aligned}
N^2\mu(A)^2 &\le \sum_{n=1}^{N}\sum_{m=1}^{N} \mu(T^{-n}(A)\cap T^{-m}(A)) \\
&= N\mu(A) + 2\sum_{k=1}^{N-1}(N-k)\mu(A\cap T^{-k}(A)),
\end{aligned}$$

or, after dividing by N^2,

$$\mu(A)^2 \le \frac{\mu(A)}{N} + \frac{1}{N}\sum_{k=1}^{N-1}\left(1-\frac{k}{N}\right)\mu(A\cap T^{-k}(A)).$$

In other words, a weighted average of the $\mu(A \cap T^{-k}(A))$'s is at least $\mu(A)^2$. It follows that for any $\varepsilon > 0$, there must be infinitely many values of k such that $\mu(A \cap T^{-k}(A)) > \mu(A)^2 - \varepsilon$. ■

5.3 A First Look at Multiple Recurrence

Let A be a set of positive measure in a finite measure space. Then we know there is some positive integer n_1 so that $\mu(A \cap T^{-n_1}(A)) > 0$. So if we let $B = A \cap T^{-n_1}(A)$, we know there is some positive integer n_2 so that $\mu(B \cap T^{-n_2}(B)) > 0$. If we express everything in terms of A, that means that

$$\begin{aligned}
\mu(B\cap T^{-n_2}(B)) &= \mu((A\cap T^{-n_1}(A))\cap T^{-n_2}(A\cap T^{-n_1}(A))) \\
&= \mu(A\cap T^{-n_1}(A)\cap T^{-n_2}(A)\cap T^{-n_1-n_2}(A)) \\
&> 0.
\end{aligned}$$

It would be nice if we could take $n_2 = n_1$, so as to say that

$$\mu(A \cap T^{-n_1}(A) \cap T^{-2n_1}(A)) > 0.$$

However, there's nothing in Poincaré recurrence that permits this simplification.

However, the result is still true! In fact, Furstenberg proved a great generalization of Poincaré recurrence, known as the Furstenberg Multiple Recurrence Theorem.

Theorem 5.7 (Furstenberg Multiple Recurrence Theorem [**Fur77**]). *Let $(X, \mathcal{A}, \mu, T)$ be a measure-preserving system. If $A \in \mathcal{A}$ with $\mu(A) > 0$, then for every positive integer k, there is a positive integer n (depending on k and A) such that*

$$\mu(A \cap T^{-n}(A) \cap T^{-2n}(A) \cap \cdots \cap T^{-(k-1)n}(A)) > 0.$$

We will not be able to give a proof of the Furstenberg Multiple Recurrence Theorem, since it is too difficult. However, we will study some of its applications in Chapter 9. One of these applications is *Szemerédi's Theorem.* This is interesting because Szemerédi's Theorem doesn't appear to have anything to do with ergodic theory or measure-preserving transformations.

Theorem 5.8 (Szemerédi, [**Sze75**]). *If A is any subset of the positive integers with positive upper density, then A contains arbitrarily long arithmetic progressions.*

What it means for a set A to have *positive upper density* is that there exists an $\varepsilon > 0$ such that there are infinitely many positive integers N such that

$$\frac{|A \cap \{1, 2, \ldots, N\}|}{N} > \varepsilon.$$

A *k-term arithmetic progression* in A is a sequence of the form $(a, a + d, a + 2d, \ldots, a + (k-1)d)$ of elements of A, where $d > 0$. Szemerédi's Theorem says that a set with positive upper density has k-term arithmetic progressions for all positive integers k.

In Chapter 9, we will see how to use the Furstenberg Multiple Recurrence Theorem to prove Szemerédi's Theorem, and vice versa.

5.4 Invariant Sets and Ergodicity

Definition 5.9. Let $(X, \mathcal{A}, \mu)$ be a measure space and T a measure-preserving transformation. A set $A \in \mathcal{A}$ is said to be *positively* T*-invariant* if whenever $x \in A$, then $T(x) \in A$ as well.

Actually, we'll be a little less strict about our definition. As with so many things in measure theory and ergodic theory, we wish to discount null sets. So, with that in mind, let's modify the above definition slightly:

Definition 5.10. Let $(X, \mathcal{A}, \mu)$ be a measure space and T a measure-preserving transformation. A set $A \in \mathcal{A}$ is said to be *positively* T*-invariant* if $\mu(A \setminus T^{-1}(A)) = 0$.

In other words, if $x \in A$, then $T(x) \in A$ as well, except on a subset of A of measure 0.

Often we would like not only that if $x \in A$, then $T(x) \in A$, but also the converse: if $T(x) \in A$, then $x \in A$, again except on a null set.

Definition 5.11. Let $(X, \mathcal{A}, \mu)$ be a measure space and T a measure-preserving transformation. A set $A \in \mathcal{A}$ is said to be T*-invariant* if $\mu(A \triangle T^{-1}(A)) = 0$.

The triangle denotes the symmetric difference:

$$A \triangle B = (A \setminus B) \cup (B \setminus A).$$

The notions of T-invariance and positive T-invariance coincide when $(X, \mathcal{A}, \mu)$ is a finite measure space; see Problem 1.

In any measure-preserving transformation on a finite measure space, we have some obvious T-invariant subsets of X. In particular, any subset of X with measure 0 is T-invariant for any T, as is the complement of any measure-0 subset of X. Whether or not there are other T-invariant subsets of X is an interesting property of the measure space $(X, \mathcal{A}, \mu)$ and the transformation T.

Definition 5.12. A measure-preserving transformation T on a measure space $(X, \mathcal{A}, \mu)$ is said to be *ergodic* if whenever $A \in \mathcal{A}$ is a T-invariant set, then either $\mu(A) = 0$ or $\mu(A^c) = 0$. If $(X, \mathcal{A}, \mu, T)$ is a measure-preserving system and T is ergodic, we say that $(X, \mathcal{A}, \mu, T)$ is an *ergodic system.*

If T is not ergodic, then we can think of it as decomposing as separate transformations on almost disjoint sets. For instance, consider the rotation $R_{\frac{1}{2}}$ on $\mathbb{R}/\mathbb{Z}$, defined by $R_{\frac{1}{2}}(x) = x + \frac{1}{2} \pmod 1$. Let $A = \left[0, \frac{1}{4}\right) \cup \left[\frac{1}{2}, \frac{3}{4}\right)$. Then A is $R_{\frac{1}{2}}$-invariant, as is $B = A^c = \left[\frac{1}{4}, \frac{1}{2}\right) \cup \left[\frac{3}{4}, 1\right)$. That is, $T(A) = A$ and $T(B) = B$. So T acts entirely separately on A and B, and the two subsets never interact as far as T is concerned. The ergodic transformations are not decomposable in this way.

Ergodicity is quite distinct from recurrence, in that there are transformations that are ergodic but not recurrent, as well as transformations that are recurrent but not ergodic. For example, $R_{\frac{1}{2}}$ is recurrent but not ergodic, whereas $T : \mathbb{Z} \to \mathbb{Z}$ defined by $T(x) = x + 1$ is ergodic but not recurrent.

Suppose that $(X, \mathcal{A}, \mu, T)$ is a measure-preserving system. We know that if $A \in \mathcal{A}$ is any measurable set with positive measure, then for almost all $x \in A$, some T-iterate of x eventually lands back in A. But what happens if $x \notin A$? If T is not ergodic, then it might be the case that no iterate of x ever lands in A. For instance, if $T = R_{\frac{1}{2}}$ on $\mathbb{R}/\mathbb{Z}$, $A = \left[0, \frac{1}{4}\right) \cup \left[\frac{1}{2}, \frac{3}{4}\right)$ and $x = \frac{3}{8}$, then the iterates of x are $\frac{3}{8}, \frac{7}{8}, \frac{3}{8}, \frac{7}{8}, \dots$, which never land in A. More generally, if $x \in A^c = \left[\frac{1}{4}, \frac{1}{2}\right) \cup \left[\frac{3}{4}, 1\right)$, then no iterate of x ever lands in A. And this is because A and A^c are T-invariant. However, if T is ergodic, then this sort of thing doesn't happen. In Chapter 6, we will see some examples of ergodic transformations on $\mathbb{R}/\mathbb{Z}$, and what the consequence is of that. As we shall see, checking that a transformation is ergodic is not a trivial task.

Proposition 5.13. *Let $(X, \mathcal{A}, \mu, T)$ be an ergodic system. Let $A \in \mathcal{A}$ be any measurable set with $\mu(A) > 0$. Then for almost all $x \in X$, there is some positive integer n such that $T^n(x) \in A$.*

Proof. Let $B = \{x \in X : T^n(x) \notin A \text{ for all nonnegative integers } n\}$. We will show that B is T-invariant, i.e. that $\mu(B \vartriangle T^{-1}(B)) = 0$. First, suppose that $x \in B \setminus T^{-1}(B)$. That means that $x \in B$ but $T(x) \notin B$, i.e. that no iterate of x lies in A, but some iterate of $T(x)$ lies in A. This is impossible, as any iterate of $T(x)$ is also an iterate of x. Thus $B \setminus T^{-1}(B) = \emptyset$. Now suppose that $x \in T^{-1}(B) \setminus B$. This means that $T(x) \in B$ but $x \notin B$, i.e. that no iterate of $T(x)$ lies in A, but some iterate of x lies in A.

The only iterate of x that isn't an iterate of $T(x)$ is x itself. This means that $x \in A$, but x never returns to A. Since X is assumed to be finite and hence recurrent, it follows that the set of such x is a null set. Thus we have $\mu(T^{-1}(B) \setminus B) = 0$, so B is T-invariant. Since T is assumed to be ergodic, either $\mu(B) = 0$ or $\mu(B^c) = 0$. But almost all $x \in A$ already lie in B^c, so $\mu(B^c) \geq \mu(A) > 0$. Thus it must be the case that $\mu(B) = 0$, so for almost all $x \in X$, some iterate of x lies in A. ∎

In the Poincaré Recurrence Theorem, we could ask how long it takes for a point $x \in A$ to return to A. That is, what is the smallest positive integer $n(x)$ such that $T^{n(x)}(x) \in A$? Intuitively, it seems that $n(x)$ will tend to be smaller when T is not ergodic than when T is ergodic, because in the non-ergodic case, the iterates of x are more restricted. Indeed, this is true. Here is a quantitative statement regarding the amount of time it takes to return.

Theorem 5.14 (Kac's Lemma, [**Kac47**]). *Let $(X, \mathcal{A}, \mu, T)$ be an ergodic system. Let $A \in \mathcal{A}$ have positive measure. For each $x \in A$, let $n_A(x)$ be the smallest positive integer n such that $T^n(x) \in A$; $n(x)$ is finite except on a null set. We have*

$$\int_A n_A \, d\mu = \mu(X).$$

I don't know if Kac's Lemma is exactly intuitively clear, but it seems plausible at least: if we have a small set A, then n_A will tend to be pretty large, whereas if we have a large set B, then n_B will tend to be pretty small. But in the former case, we're integrating over a small set A, whereas in the latter case, we're integrating over a larger set B. So the integral of a large function over a small set could conceivably be equal to the integral of a small function over a large set. And indeed they are!

Proof. For each positive integer n, let

$$\begin{aligned}
A_n &= A \cap T^{-1}(A^c) \cap T^{-2}(A^c) \cap \cdots \cap T^{-(n-1)}(A^c) \cap T^{-n}(A), \\
B_n &= A^c \cap T^{-1}(A^c) \cap T^{-2}(A^c) \cap \cdots \cap T^{-(n-1)}(A^c) \cap T^{-n}(A),
\end{aligned}$$

so A_n consists of those points that start in A and return to A for the first time after n iterates, while B_n consists of those points that do not start in A but reach A for the first time after n iterates. Since T is assumed to be ergodic, we know that

$$\mu\left(\bigcap_{n=1}^{\infty}(A_n^c \cap B_n^c)\right) = 0,$$

i.e. almost all points in X eventually reach A. Furthermore, all the A_n's and B_n's are disjoint, so except for a null set, they partition X. Thus we have

$$\mu(X) = \sum_{n=1}^{\infty}(\mu(A_n) + \mu(B_n)). \tag{5.1}$$

Now, observe that

$$T^{-1}(B_n) = A_{n+1} \cup B_{n+1},$$

because $x \in T^{-1}(B_n)$ means that $T(x) \notin A$ and then $T(x)$ does not return to A for another n iterates. If $x \in A$, then this means that $x \in A_{n+1}$, while if $x \notin A$, then $x \in B_{n+1}$.

Since T is measure-preserving, we have

$$\mu(B_n) = \mu(T^{-1}(B_n)) = \mu(A_{n+1}) + \mu(B_{n+1}).$$

Applying this relation repeatedly, we find that for every $m > n$, we have

$$\mu(B_n) = \mu(B_m) + \sum_{i=n+1}^{m} \mu(A_i).$$

Since $\mu(X) < \infty$, (5.1) implies that $\lim_{m\to\infty} \mu(B_m) = 0$. Thus we have

$$\mu(B_n) = \sum_{i=n+1}^{\infty} \mu(A_i).$$

Substituting this into (5.1), we get

$$\begin{aligned}\mu(X) &= \sum_{n=1}^{\infty}\Big(\mu(A_n) + \sum_{i=n+1}^{\infty}\mu(A_i)\Big)\\ &= \sum_{n=1}^{\infty}\sum_{i=n}^{\infty}\mu(A_i)\\ &= \sum_{i=1}^{\infty}\sum_{n=1}^{i}\mu(A_i)\\ &= \sum_{i=1}^{\infty} i\mu(A_i)\\ &= \int_A n_A \, d\mu,\end{aligned}$$

as desired. ■

5.5 Induced Transformations

Let $(X, \mathcal{A}, \mu, T)$ be a measure-preserving system. Let $A \in \mathcal{A}$ have positive measure. Now, for almost all $x \in A$, there is some $n_A(x)$ such that $T^{n_A(x)}(x) \in A$; choose $n_A(x)$ to be the minimal positive integer with this property. Let us remove the measure-0 subset of A on which no iterate lies in A, as well as all points whose iterates eventually lie in this measure-0 subset; that set still has measure 0. So we may assume that for *all* $x \in A$, there is such a positive integer $n_A(x)$. We can now define a new transformation, just on A.

Definition 5.15. With hypotheses as above, the *induced transformation* on A is $T_A : A \to A$ defined by $T_A(x) = T^{n_A(x)}(x)$.

In other words, T_A just keeps applying T until we return to A. It isn't immediately obvious that T_A is measure-preserving on A, but this turns out to be true.

Proposition 5.16. *The induced transformation T_A is a measure-preserving transformation on A.*

Proof. The proof should be somewhat reminiscent of that of Kac's Lemma. For each positive integer n, let $A_n = \{x \in A : n_A(x) = n\}$, and let $B_n = A^c \cap T^{-1}(A)^c \cap \cdots \cap T^{-(n-1)}(A)^c \cap T^{-n}(A)$. That is, B_n is the set of points in $X \backslash A$ such that the amount of time needed to reach A is exactly n. Note that $T^{-1}(A) = A_1 \sqcup B_1$ and $T^{-1}(B_n) = A_{n+1} \cup B_{n+1}$. For any measurable subset $C \in \mathcal{A}$ such that $C \subseteq A$, we have $\mu(C) = \mu(T^{-1}(C))$ since T is measure-preserving. We need to show that $\mu(C) = \mu(T_A^{-1}(C))$ in order to show that T_A is measure-preserving, so to do this, we'll show that $\mu(T^{-1}(C)) = \mu(T_A^{-1}(C))$. Since the A_n's cover A, we have

$$T_A^{-1}(C) = \bigcup_{n=1}^{\infty} A_n \cap T_A^{-1}(C) = \bigcup_{n=1}^{\infty} A_n \cap T^{-n}(C),$$

and since the A_n's are disjoint, we have

$$\mu(T_A^{-1}(C)) = \sum_{n=1}^{\infty} \mu(A_n \cap T^{-n}(C)).$$

We also have

$$\begin{aligned}
\mu(T^{-1}(C)) &= \mu(A_1 \cap T^{-1}(C)) + \mu(B_1 \cap T^{-1}(C)) \\
&= \mu(A_1 \cap T^{-1}(C)) + \mu(T^{-1}(B_1 \cap T^{-1}(C))) \\
&= \mu(A_1 \cap T^{-1}(C)) + \mu(A_2 \cap T^{-2}(C)) + \mu(B_2 \cap T^{-2}(C)) \\
&= \cdots \\
&= \sum_{k=1}^{n} \mu(A_k \cap T^{-k}(C)) + \mu(B_n \cap T^{-n}(C)).
\end{aligned}$$

Since

$$\mu(X) \geq \mu\left(\bigcup_{n=1}^{\infty} B_n \cap T^{-n}(C)\right) = \sum_{n=1}^{\infty} \mu(B_n \cap T^{-n}(C)),$$

we have

$$\lim_{n\to\infty} \mu(B_n \cap T^{-n}(C)) = 0.$$

Thus we have

$$\mu(C) = \mu(T^{-1}(C)) = \sum_{k=1}^{\infty} \mu(A_k \cap T^{-k}(C)) = \mu(T_A^{-1}(C)),$$

as desired. ∎

5.6 Problems

(1) Prove that if $(X, \mathcal{A}, \mu, T)$ is a measure-preserving system, then a set $A \in \mathcal{A}$ is T-invariant if and only if it is positively T-invariant. Prove that the notions of T-invariance and positive T-invariance do not necessarily coincide when $\mu(X) = \infty$.

(2) Let $([0,1], \mathcal{L}, \lambda, T)$ be a measure-preserving system. Prove that for almost every $x \in [0,1]$, there is a sequence $n_1 < n_2 < \cdots$ of positive integers such that $\lim_{k\to\infty} T^{n_k}(x) = x$.

(3) Let $(X, \mathcal{A}, \mu, T)$ be a measure-preserving system. Suppose that $A \in \mathcal{A}$ and $\mu(A) > 0$. Let $n = n(A)$ be the smallest positive integer such that $\mu(T^{-n}(A) \cap A) > 0$. Find the smallest positive integer $b(r)$ such that $n(A) \leq b(r)$ whenever $\mu(A) \geq r$. Prove that your number is optimal by giving an example where $n(A) = b(r)$ for every $r \in [0,1]$. Note that $b(r)$ should depend only on r, not on T or X.

(4) Let A denote the set of $x \in [0,1)$ with the following property: Suppose that the decimal expansion of x is $0.a_1a_2a_3\cdots$, where each $a_i \in \{0,1,2,\ldots,9\}$. Then A denotes the set of those x such that for all positive integers k, the string $a_1a_2\cdots a_k$ occurs infinitely often as consecutive digits of x. (For example, if $x = .3752663\cdots$, this means that the decimal expansion of x contains infinitely many 3's, infinitely many 37's, infinitely many 375's, and so forth.) Prove that $\mu(A) = 1$.

(5) Prove with bare hands that if $\alpha \in \mathbb{R}$, then the rotation R_α on $\mathbb{R}/\mathbb{Z}$ satisfies multiple recurrence (i.e. that the Furstenberg Multiple Recurrence Theorem holds for R_α).

(6) A set P of positive integers is said to be a *Poincaré sequence* if for every measure-preserving system $(X, \mathcal{A}, \mu, T)$ and any set $A \in \mathcal{A}$ with $\mu(A) > 0$, there is some $n \in P$ such that $\mu(T^{-n}(A) \cap A) > 0$. Show that if S is any infinite set of positive integers, then $P = \{s - t : s, t \in S, s > t\}$ is a Poincaré sequence.

(7) A set Q of positive integers is said to be a *thick set* if for every positive integer k, there exist k consecutive integers in Q. Prove that a thick set is a Poincaré sequence.

(8) Let $(X, \mathcal{A}, \mu)$ be a measure space and T a measure-preserving transformation. A measurable set $W \in \mathcal{A}$ is said to be a *wandering set* for T if for all $i, j \geq 0$ with $i \neq j$, $T^{-i}(W) \cap T^{-j}(W) = \emptyset$. Prove that T is recurrent if and only if it contains no wandering sets of positive measure.

(9) Let $(X, \mathcal{A}, \mu)$ be a measure space, and let T be an invertible measure-preserving transformation. Let $B \in \mathcal{A}$ have finite (possibly 0) measure. Prove that for any measurable set E with $\mu(E) > 0$ (possibly $\mu(E) = \infty$), almost every $x \in E$ either returns to E infinitely often, or it only has finitely many iterates in B.

(10) We say that a measure space $(X, \mathcal{A}, \mu)$ is *σ-finite* if there exist measurable subsets $X_1, X_2, \ldots$ of X with $\mu(X_n) < \infty$ for all n, $X_1 \subseteq X_2 \subseteq \cdots$, and $X = \bigcup_{n=1}^{\infty} X_n$. Let T be an invertible measure-preserving transformation on a σ-finite measure space $(X, \mathcal{A}, \mu)$. We say that a point $x \in X$ *goes to* ∞ if for each X_n, there are only finitely many T-iterates of x lying in X_n. Show that, if $A \in \mathcal{A}$ is any measurable set with $\mu(A) > 0$, almost every point $x \in A$ either returns to A an infinite number of times or goes to ∞.

(11) Let $(X, \mathcal{A}, \mu, T)$ be a measure-preserving system. Let $A \in \mathcal{A}$ have $\mu(A) > 0$. Prove that almost every $x \in A$ spends a *positive fraction* of time in A, i.e.

$$\limsup_{n \to \infty} \frac{\#\{0 \leq j \leq n-1 : T^j(x) \in A\}}{n} > 0.$$

(This is also true with lim inf instead of lim sup, but this is more difficult to prove.)

(12) A sequence $n_1 < n_2 < n_3 < \cdots$ of positive integers is said to be *syndetic* if there is some positive integer C such that $n_{i+1} - n_i \leq C$ for all i. The goal of this problem is to prove that if $(X, \mathcal{A}, \mu, T)$ is a measure-preserving system, and $A \in \mathcal{A}$ with $\mu(A) > 0$, then the sequence of numbers n such that $\mu(A \cap T^{-n}(A)) > 0$ is syndetic.

(a) Note that for any increasing sequence $k_1 < k_2 < \cdots$, there exist $j > i \geq 1$ such that $\mu(A \cap T^{-(k_j - k_i)}(A)) > 0$. (This is Problem 6, so go solve that problem if you haven't already.)

(b) Given any infinite sequence $K = (k_1, k_2, \ldots)$ of positive integers (not necessarily increasing), let S_K denote the set of all sums of consecutive terms of K, i.e. $S_K = \{k_i + k_{i+1} + \cdots + k_j : i \leq j\}$. Prove that S_K contains some n such that $\mu(A \cap T^{-n}(A)) > 0$.

(c) Prove that $\{n : \mu(A \cap T^{-n}(A))\}$ is syndetic.

(13) Let $([0,1], \mathcal{L}, \lambda, T)$ be a measure-preserving system. Prove that for almost all $x \in [0,1]$,

$$\liminf_{n\to\infty} n|T^n(x) - x| \leq 1. \tag{5.2}$$

(14) Let $T : [0,1) \to [0,1)$ be defined by $T(x) = x + \phi - \lfloor x + \phi \rfloor$, where $\phi = \frac{1+\sqrt{5}}{2}$.

(a) Check that T is measure-preserving with respect to the Lebesgue measure λ.

(b) Let $q_n = \frac{\lfloor x + n\phi \rfloor}{n}$. Note that $\lim_{n\to\infty} q_n = \phi$ for all $x \in [0,1)$. Check that

$$n|T^n(x) - x| = n^2|\phi - q_n|$$

for all n.

(c) Prove that

$$\liminf_{n\to\infty} n^2|\phi - q_n| \geq \frac{1}{\sqrt{5}}.$$

(d) Show that the constant 1 on the right side of (5.2) of Problem 13 cannot be replaced by anything smaller than $\frac{1}{\sqrt{5}}$.

(e) What is the best possible constant on the right side of (5.2)?

Chapter 6

Ergodicity

6.1 Proving Ergodicity

Recall from Chapter 5 the definition of ergodicity: a measure-preserving transformation T on $(X, \mathcal{A}, \mu)$ is ergodic if whenever $A \in \mathcal{A}$ is T-invariant, then either $\mu(A) = 0$ or $\mu(A^c) = 0$. It isn't so straightforward to check that a transformation is ergodic, since this requires determining all the invariant sets somehow. Here is a lemma that is useful for this purpose:

Lemma 6.1 (Knopp). *Let $B \subseteq \mathbb{R}/\mathbb{Z}$ be a Lebesgue-measurable set, and let $\mathcal{C}$ be a set of subintervals of $[0, 1)$ satisfying the following two properties:*

(1) *Every open subinterval (a, b) of $[0, 1)$ is a finite or countable union of disjoint sets in $\mathcal{C}$, and*

(2) *There exists a real number $\gamma > 0$ such that for every $A \in \mathcal{C}$, $\lambda(A \cap B) \geq \gamma\lambda(A)$.*

Then $\lambda(B) = 1$.

In other words, if a set takes up a bounded-away-from-zero proportion of each of enough subintervals, then it must have full measure.

Proof. Suppose, on the contrary, that $\lambda(B^c) > 0$. Since B is Lebesgue measurable, we may write $B = C \cup D$, where C is a Borel set and D is a null set, and $\lambda(B) = \lambda(C)$. Thus if $\lambda(B^c) > 0$, then $\lambda(C^c) > 0$. Now, for any $\varepsilon > 0$, there is a set E_ε which is a finite disjoint union of open intervals such that $\lambda(C^c \vartriangle E_\varepsilon) < \varepsilon$. (This was Problem 5 from Chapter 2.)

By the conditions, E_ε is a countable union of disjoint elements of $\mathcal{C}$, so $\lambda(C \cap E_\varepsilon) \geq \gamma\lambda(E_\varepsilon)$.

Now, since

$$\lambda(C^c \triangle E_\varepsilon) \geq \lambda(C \cap E_\varepsilon) \geq \gamma\lambda(E_\varepsilon) \geq \gamma\lambda(C^c \cap E_\varepsilon) > \gamma(\lambda(C^c) - \varepsilon),$$

we have

$$\gamma(\lambda(C^c) - \varepsilon) < \lambda(C^c \triangle E_\varepsilon) < \varepsilon.$$

Thus $\gamma\lambda(C^c) < \varepsilon + \gamma\varepsilon$. This is true for any $\varepsilon > 0$, so $\gamma\lambda(C^c) = 0$, and $\lambda(C^c) = 0$. But this contradicts our assumption on $\lambda(B^c)$, and hence $\lambda(C^c)$, being nonzero. ■

Let's see how to use Knopp's Lemma to prove that certain transformations are ergodic. Let us start by showing that the transformation T on $\mathbb{R}/\mathbb{Z}$ given by $T(x) = 2x - \lfloor 2x \rfloor$ is ergodic; the same argument will work just as well for $T(x) = bx - \lfloor bx \rfloor$, but for concreteness, we'll just work with $b = 2$. We take $\mathcal{C}$ to be the set of all intervals of the form $\left(\frac{a}{2^n}, \frac{a+1}{2^n}\right)$ for some a and n.

Suppose now that B is a T-invariant set with positive measure. Let $E = \left(\frac{a}{2^n}, \frac{a+1}{2^n}\right)$ be an element of $\mathcal{C}$. Then we have

$$\frac{\lambda(B \cap E)}{\lambda(E)} = \frac{\lambda(T^{-n}(B) \cap E)}{\lambda(E)} = \frac{\lambda(B \cap T^n(E))}{\lambda(T^n(E))} = \lambda(B), \tag{6.1}$$

since $T^n(E)$ is all of $\mathbb{R}/\mathbb{Z}$, up to a null set. Thus $\lambda(B \cap E) = \lambda(B)\lambda(E)$ for every $E \in \mathcal{C}$. If we let $\gamma = \lambda(B)$, then the hypotheses of Knopp's Lemma apply, so we have $\lambda(B) = 1$. Thus we may conclude that T is ergodic.

By a similar argument, we can show that Bernoulli shifts are ergodic. Let $(X, \mathcal{A}, \mu)$ be a probability space, and let $(Y, \mathcal{B}, \nu)$ be a new probability space, with $Y = X^{\mathbb{N}}$ the space of all infinite sequences of elements of X, i.e. $Y = \{(x_1, x_2, \ldots) : x_i \in X\}$. The σ-algebra $\mathcal{B}$ on Y is the product σ-algebra, i.e. the smallest σ-algebra containing all sets of the form $A_1 \times A_2 \times \cdots \times A_n \times X \times X \times \cdots$, where each $A_i \in \mathcal{A}$. The measure ν is the product measure on Y, i.e. $\nu(A_1 \times A_2 \times \cdots \times A_n \times X \times X \times \cdots) = \prod_{i=1}^n \mu(A_i)$; this is enough to define a measure on all of $\mathcal{B}$ because all sets in $\mathcal{B}$ can be expressed in terms of these "rectangles." Let $T : Y \to Y$ be the function $T(x_1, x_2, \ldots) = (x_2, x_3, \ldots)$. This is a measure-preserving transformation, because it's a Bernoulli shift, and we have already seen in Chapter 4 that Bernoulli shifts are measure-preserving. We will now see that it is also ergodic. To do this, we need to replace Knopp's Lemma by a version better suited for Y. For this, let $\mathcal{C}$ be the set of cylinder sets in Y, i.e. sets of the form $A_1 \times A_2 \times \cdots \times A_n \times X \times X \times \cdots$, where $A_1, \ldots, A_n \in \mathcal{A}$.

So, now take a T-invariant set $B \in \mathcal{B}$ with $\nu(B) > 0$, and let E be a cylinder set of the form $A_1 \times A_2 \times \cdots \times A_n \times X \times X \times \cdots$ with positive measure. Then (6.1) holds for this B and E as well, and the rest of the argument goes through in exactly the same way. Thus T is ergodic.

You will show in Problem 5 that irrational rotations are ergodic.

6.2 The Strong Law of Large Numbers

Before we discuss the ergodic theorem, we'll start by seeing what it generalizes: the Strong Law of Large Numbers. This is a very important theorem in probability theory. Here is the statement:

Theorem 6.2 (Strong Law of Large Numbers). *Let $(X, \mathcal{A}, \mu)$ be a probability space. Let $X_1, X_2, \ldots$ be independent and identically distributed random variables on $(X, \mathcal{A}, \mu)$, with $|\mathbb{E}(X_i)| < \infty$. Then*

$$\lim_{n\to\infty} \frac{X_1 + \cdots + X_n}{n} = \mathbb{E}(X_1)$$

almost surely.

While this theorem likely contains a bunch of unfamiliar words, it probably doesn't really contain any unfamiliar *ideas*. For example, consider rolling a normal six-sided die. Then $X = \{1, 2, 3, 4, 5, 6\}$, the possible outcomes. We can take $\mathcal{A} = \mathcal{P}(X)$ and μ to be the uniform measure: $\mu(1) = \mu(2) = \cdots = \mu(6) = \frac{1}{6}$. A *random variable* is essentially a function from a measure space to the real numbers. In this case, we consider the random variable to be essentially the identity function. So, this means that $X_1, X_2, \ldots$ are the results of independent die rolls. (This is in contrast to *dependent* die rolls. For example, we could roll a die once, and then let $X_1, X_2, \ldots$ all be the result of that single die roll. The Strong Law of Large Numbers doesn't necessarily hold for dependent random variables.) Now, $\mathbb{E}(X_i)$ denotes the *expected value*, or simply the average (or arithmetic mean). In this case, that's $\frac{1+2+3+4+5+6}{6} = \frac{7}{2}$. Note that $\mathbb{E}(X_i)$ is the same for all i, so we may freely replace $\mathbb{E}(X_i)$ with $\mathbb{E}(X_1)$ or something else similar when convenient.

Putting all this together, this says that if we roll a die many times and average the results of the individual rolls together, we will get a result very close to $\frac{7}{2}$. We have this phrase "almost surely" in the statement of the theorem. "Almost surely" is basically the same as "almost everywhere," i.e. the set of sequences where the limit of the averages isn't equal to the mean is a null set. For example, in the die roll case, there certainly are sequences where this fails: perhaps all the rolls are 1's, for instance. So that's why we need "almost surely" as part of the theorem statement.

I think of the Strong Law of Large Numbers as being less a theorem and more a sanity check: if the average of independent observations *weren't* equal to the mean, that would only imply that we've asked the wrong question and should go back to the drawing board and figure out what the right question is. But it is true. We won't prove it just yet, because it's a special case of the Birkhoff Ergodic Theorem, which we will state and prove shortly.

Recall that a real number α is said to be *simply normal to base b* if, in its base-b expansion, it contains roughly the same number of 0's, 1's, ..., $(b-1)$'s. More precisely, let us suppose that $\alpha \in [0,1)$, and that $\alpha = 0.a_1a_2a_3\cdots$ in base b. Then α is simply normal to base b if, for all $d \in \{0,1,\dots,b-1\}$, we have

$$\lim_{n\to\infty} \frac{\#\{i \in \{1,2,\dots,n\} : a_i = d\}}{n} = \frac{1}{b}.$$

Theorem 6.3. *For any integer $b \geq 2$, almost all $\alpha \in [0,1)$ are simply normal to base b.*

Proof. Pick a random number $\alpha \in [0,1)$, and fix a digit $d \in \{0,1,\dots,b-1\}$. Define random variables $X_1, X_2, \dots$ by

$$X_i(\alpha) = \begin{cases} 1 & a_i(\alpha) = d, \\ 0 & a_i(\alpha) \neq d. \end{cases}$$

The digits of α are independent and identically distributed, so the X_i's are independent and identically distributed random variables, with mean $\frac{1}{b}$. Thus the Strong Law of Large Numbers implies that

$$\lim_{n\to\infty} \frac{X_1 + \cdots + X_n}{n} = \frac{1}{b}.$$

But $X_1 + \cdots + X_n$ is just the number of the first n digits that are equal to d. ■

We'd really like to prove a stronger statement here: not only are there roughly the same number of 0's, 1's, ..., $(b-1)$'s, but there are also roughly the same number of each possible sequence of length 2, or of length 3, and so forth.

Definition 6.4. A number $\alpha \in [0,1)$ is said to be *normal* to base b if for any positive integer m and any sequence of digits $d_1, \dots, d_m$ of length m with $d_i \in \{0, 1, \dots, b-1\}$, we have

$$\lim_{n\to\infty} \frac{\#\{i \in \{1,2,\dots,n\} : b_i = d_1, b_{i+1} = d_2, \dots, b_{i+m-1} = d_m\}}{n} = \frac{1}{b^m}.$$

We might hope to use the Strong Law of Large Numbers on blocks of length 2, but this doesn't work. The reason is that these blocks are not independent. Suppose that $b = 10$, for instance, and we consider the block 13. For $\alpha \in [0,1)$, we could try defining random variables $X_i(\alpha)$ by

$$X_i(\alpha) = \begin{cases} 1 & a_i(\alpha) = 1, a_{i+1}(\alpha) = 3, \\ 0 & \text{otherwise.} \end{cases}$$

But the problem is that these $X_i(\alpha)$'s are *not* independent. For instance, if $X_i = 1$, then X_{i+1} *must* be 0, because we can't have two overlapping 13-blocks.

To deal with this, we need a generalization of the Strong Law of Large Numbers that doesn't require full independence, but only something weaker. And that is what the ergodic theorem will give us.

6.3 The Birkhoff Ergodic Theorem

Theorem 6.5 (Birkhoff, [**Bir31**]). *Let $(X, \mathcal{A}, \mu, T)$ be a measure-preserving system. Then for any $f \in L^1(\mu)$, we have*

$$\lim_{n\to\infty} \frac{1}{n} \sum_{i=0}^{n-1} f \circ T^i(x) = f^*(x)$$

exists for almost all $x \in X$. Furthermore, the function f^* is T-invariant in the sense that $f^*(x) = f^*(T(x))$ for almost all x, and

$$\int_X f\, d\mu = \int_X f^*\, d\mu.$$

If furthermore T is ergodic, then f^* is constant almost everywhere, and

$$f^* = \int_X f\, d\mu$$

almost everywhere.

Before we turn to the proof of the ergodic theorem, let's see some of the things we can do with it. First, we can prove the Strong Law of Large Numbers. This isn't so obvious, because the Strong Law of Large Numbers is about independent random variables, whereas the ergodic theorem is about iterates of a single element. However, we can get around this issue by considering our space to be infinite sequences of observations, with the map T being a generalized version of a Bernoulli shift.

Proof of the Strong Law of Large Numbers. Let $(X, \mathcal{A}, \mu)$ be a probability space, as in the statement of the Strong Law of Large Numbers, and let $(Y, \mathcal{B}, \nu)$ be the product space, where $Y = X^{\mathbb{N}}$, and $\mathcal{B}$ and ν are as in §6.1. As we have already seen, the shift operator $T : Y \to Y$ given by $T(x_1, x_2, \ldots) = (x_2, x_3, \ldots)$ is ergodic. For a sequence $x = (x_1, x_2, \ldots) \in Y$, let $f(x)$ be the random variable associated with x_1. Thus $f(x) = X_1$, $f(T(x)) = X_2$, and so forth. Then the Birkhoff Ergodic Theorem says that

$$\lim_{n\to\infty} \frac{1}{n} \sum_{i=0}^{n-1} f(T^i(x)) = \int_X f\, d\mu$$

almost surely. The right side is equal to $\mathbb{E}(X_1) = \mathbb{E}(X_2) = \cdots$, since that's the definition of expected value. The left side is simply

$$\lim_{n\to\infty} \frac{X_1 + \cdots + X_n}{n}.$$

Thus we have

$$\lim_{n\to\infty} \frac{X_1 + \cdots + X_n}{n} = \mathbb{E}(X_1)$$

almost surely. ∎

Another thing we can do with the ergodic theorem is to prove that almost all numbers are normal. To simplify notation a bit, let's look at the case $b = 2$, where we are looking for appearances of the subsequence 01, but the same proof will work in complete generality, in any base, for any string of arbitrary length, to prove the following theorem:

Theorem 6.6 (Borel, [**Bor09**]). *For any base b, almost all real numbers are normal to base b.*

To do this in the case of $b = 2$ and the string 01, let's define a function $f : \mathbb{R}/\mathbb{Z} \to \mathbb{R}$ by setting $f(\alpha) = 1$ if the first two binary digits of α after the decimal place are 01, and $f(\alpha) = 0$ otherwise; in other words, if $\alpha \in \left[\frac{1}{4}, \frac{1}{2}\right)$. We let T be the function $T(x) = 2x - \lfloor 2x \rfloor$. Since T is ergodic as we have already seen, the ergodic theorem tells us that $f^*(x)$, defined by

$$f^*(x) = \lim_{n\to\infty} \frac{1}{n} \sum_{i=0}^{n-1} f \circ T^i(x),$$

is constant almost everywhere and equal to

$$f^*(x) = \int_X f \, d\lambda = \lambda\left(\{\alpha \in \mathbb{R}/\mathbb{Z} : f(\alpha) = 1\}\right) = \lambda\left(\left[\frac{1}{4}, \frac{1}{2}\right)\right) = \frac{1}{4}.$$

In other words, for almost all real numbers $\alpha \in \mathbb{R}/\mathbb{Z}$, we get a 01-window $\frac{1}{4}$ of the time. That, together with the analogous results for other windows, is just what it means to be normal.

In Problem 4, you will do a similar analysis for Lüroth series expansions, rather than decimal expansions.

In fact, we can do even better as far as normality goes:

Theorem 6.7. *Almost all real numbers in $\mathbb{R}/\mathbb{Z}$ are normal to all integer bases $b \geq 2$.*

Proof. For each $b \geq 2$, let us set

$$E_b = \{\alpha \in \mathbb{R}/\mathbb{Z} : \alpha \text{ is not normal to base } b\}.$$

Then by Borel's normal number theorem, we have $\lambda(E_b) = 0$ for all b. Let $E = \bigcup_{b=2}^{\infty} E_b$. Then we have

$$\lambda(E) = \lambda\left(\bigcup_{b=2}^{\infty} E_b\right) \leq \sum_{b=2}^{\infty} \lambda(E_b) = 0,$$

so $\lambda(E) = 0$. But E is the set of numbers that fail to be normal to some base, so almost all $\alpha \in \mathbb{R}/\mathbb{Z}$ are normal. ■

One very important thing that the ergodic theorem tells us is that, if T is an ergodic transformation, then we eventually forget about our starting point, in the following sense: suppose we have two measurable subsets A and B, with $\mu(A), \mu(B) > 0$. Then, for large enough values of n, the probability of being in state B after roughly n steps is close to independent of whether we started in A or not. More precisely:

Theorem 6.8. *Let $(X, \mathcal{A}, \mu, T)$ be a measure-preserving system. Then T is ergodic if and only if, for all $A, B \in \mathcal{A}$ with $\mu(A), \mu(B) > 0$, we have*

$$\lim_{n\to\infty} \frac{1}{n} \sum_{j=0}^{n-1} \mu(A \cap T^{-j}(B)) = \mu(A)\mu(B). \tag{6.2}$$

Proof. Let us suppose that T is ergodic, and pick two measurable sets A and B with positive measure. Let us take $f = \mathbb{1}_A$, the indicator function for A, i.e. $\mathbb{1}_A(x) = 1$ if $x \in A$ and 0 if $x \notin A$. Then by the ergodic theorem, we have

$$\lim_{n\to\infty} \frac{1}{n} \sum_{j=0}^{n-1} \mathbb{1}_A(T^j(x)) = \mu(A)$$

for almost all $x \in X$. Multiplying both sides by $\mathbb{1}_B(x)$, we have

$$\lim_{n\to\infty} \frac{1}{n} \sum_{j=0}^{n-1} \mathbb{1}_A(T^j(x))\mathbb{1}_B(x) = \mu(A)\mathbb{1}_B(x)$$

for almost all $x \in X$. Since the left side is less than 1 for almost all $x \in X$, we may apply the Dominated Convergence Theorem, which tells us that

$$\begin{aligned}\lim_{n\to\infty} \int_X \frac{1}{n} \sum_{j=0}^{n-1} \mathbb{1}_A(T^j(x))\mathbb{1}_B(x)\, d\mu &= \int_X \lim_{n\to\infty} \frac{1}{n} \sum_{j=0}^{n-1} \mathbb{1}_A(T^j(x))\mathbb{1}_B(x)\, d\mu \\ &= \int_X \mu(A)\mathbb{1}_B(x)\, d\mu \\ &= \mu(A) \int_X \mathbb{1}_B(x)\, d\mu \\ &= \mu(A)\mu(B).\end{aligned}$$

On the other hand, we have

$$\begin{aligned}\int_X \frac{1}{n}\sum_{j=0}^{n-1} \mathbb{1}_A(T^j(x))\mathbb{1}_B(x)\,d\mu &= \frac{1}{n}\sum_{j=0}^{n-1}\int_X \mathbb{1}_{T^{-j}(A)\cap B}(x)\,d\mu \\ &= \frac{1}{n}\sum_{j=0}^{n-1}\mu(T^{-j}(A)\cap B).\end{aligned}$$

Thus we have proven that (6.2) holds.

Now, let us suppose that (6.2) holds for all $A, B \in \mathcal{A}$ with $\mu(A), \mu(B) > 0$. Then in particular if we let $B = A$, then we have

$$\lim_{n\to\infty}\frac{1}{n}\sum_{j=0}^{n-1}\mu(T^{-j}(A)\cap A) = \mu(A)^2.$$

If furthermore A is T-invariant, then we have $\mu(T^{-j}(A)\cap A) = \mu(A)$, since $T^{-j}(A)$ and A agree up to a null set. Thus we have

$$\frac{1}{n}\sum_{j=0}^{n-1}\mu(T^{-j}(A)\cap A) = \frac{1}{n}\sum_{j=0}^{n-1}\mu(A) = \mu(A).$$

Thus if A is T-invariant, we have $\mu(A) = \mu(A)^2$, so $\mu(A) = 0$ or 1. That implies that T is ergodic. ∎

We might hope for something even stronger: that if T is ergodic then we have

$$\lim_{n\to\infty}\mu(T^{-n}(A)\cap B) = \mu(A)\mu(B),$$

so that we forget the starting point on a single iteration rather than only over an average of many iterations. This is something called *mixing*, and we'll return to this topic in Chapter 8. However, this is too much to hope for in general. For example, consider the rotation R_α on $\mathbb{R}/\mathbb{Z}$. Let us suppose that A and B are small intervals. Then, for each n, $R_\alpha^{-n}(A)$ is just another small interval, of the same length as A. So, for most values of n, $R_\alpha^{-n}(A)\cap B = \varnothing$, but occasionally there's a fairly large overlap. On average, the size of the overlap is just $\lambda(A)\lambda(B)$, you'll rarely, if ever, find a value of n such that $\lambda(R_\alpha^{-n}(A)\cap B) \approx \lambda(A)\lambda(B)$. That's why we need the averages. Mixing is a stronger condition on a transformation than ergodicity.

6.4 Proof of the Ergodic Theorem

It's now time to turn to the proof of the Birkhoff Ergodic Theorem. There are several steps to this proof, and they involve a moderate dose of analysis. Let's first explain what the ideas are, and then we'll go through the details. For simplicity, we'll assume that f is an indicator of some measurable set; this is essentially sufficient, because we can then extend to simple functions, and then every measurable function can be arbitrarily well approximated by simple functions. Some care must be taken, but we'll sweep that under the rug.

So, let's take B to be some measurable set and let $f = \mathbb{1}_B$. Consider the functions $\frac{1}{n}\sum_{i=0}^{n-1} \mathbb{1}_B \circ T^i(x)$ for each $n \geq 1$, which counts the fraction of iterates of x that land in B, among the first n (starting from 0). This is some bounded sequence of measurable functions from X to $\mathbb{R}$, so we can consider the limits superior and inferior of these functions. Let's define

$$\overline{f}(x) = \limsup_{n\to\infty} \frac{1}{n}\sum_{i=0}^{n-1} \mathbb{1}_B \circ T^i(x),$$

$$\underline{f}(x) = \liminf_{n\to\infty} \frac{1}{n}\sum_{i=0}^{n-1} \mathbb{1}_B \circ T^i(x).$$

These are again measurable functions from X to $\mathbb{R}$. We need to show that $\overline{f}(x) = \underline{f}(x)$ for almost all $x \in X$. (Then there's some other stuff about the limiting function being T-invariant and precisely what function it is, and we'll deal with that bit later.) This is the most challenging part: showing that $\overline{f}(x) = \underline{f}(x)$ for almost all x. As is so often the case in analysis, when we want to prove that two functions are equal, we prove that each one is $\leq$ the other, and that's what we do here too. One direction is obvious: $\underline{f}(x) \leq \overline{f}(x)$, because a limit inferior of a sequence is always $\leq$ the limit superior. The other direction is harder, so let's get on with it.

The first thing we'll do is to prove the following:

Proposition 6.9. *With notation as above, $\overline{f}$ and $\underline{f}$ are T-invariant.*

Proof. What this means is that we must show that for all $x \in X$, we have $\overline{f}(T(x)) = \overline{f}(x)$, and similarly for $\underline{f}$. Let us write

$$a_n = \frac{1}{n}\sum_{i=0}^{n-1} \mathbb{1}_B \circ T^i(x), \qquad b_n = \frac{1}{n}\sum_{i=0}^{n-1} \mathbb{1}_B \circ T^i(T(x)).$$

Thus we have

$$\overline{f}(x) = \limsup_{n\to\infty} a_n, \qquad \overline{f}(T(x)) = \limsup_{n\to\infty} b_n,$$

and similarly for the limits inferior. Since $\mathbb{1}_B(x)$ and $\mathbb{1}_B(T^n(x))$ are either 0 or 1, we have $|a_n - b_n| \leq \frac{1}{n}$. It follows from Problem 3 that a_n and b_n have the same limits superior and inferior, i.e. that $\overline{f}(x) = \overline{f}(T(x))$ and $\underline{f}(x) = \underline{f}(T(x))$. ■

Next, we'll prove the following part of the ergodic theorem:

Proposition 6.10. *Under the hypotheses of the ergodic theorem, the limit*

$$\lim_{n\to\infty} \frac{1}{n}\sum_{i=0}^{n-1} \mathbb{1}_B \circ T^i(x)$$

exists for almost all $x \in X$.

Proof. Since $\underline{f}(x) \leq \overline{f}(x)$ for all x, we automatically have $\int_X \underline{f}\, d\mu \leq \int_X \overline{f}\, d\mu$. In order to show the reverse inequality for almost all x, it suffices to prove the opposite relation among the integrals: $\int_X \overline{f}\, d\mu \leq \int_X \underline{f}\, d\mu$. In fact, we want to show the following chain of inequalities:

$$\int_X \overline{f}\, d\mu \leq \int_X \mathbb{1}_B\, d\mu = \mu(B) \leq \int_X \underline{f}\, d\mu,$$

since then they all have to be equalities.

We first show that $\int_X \overline{f}\, d\mu \leq \mu(B)$. Let $\varepsilon > 0$, and let

$$S_n(x) = \frac{1}{n}\sum_{i=0}^{n-1} \mathbb{1}_B \circ T^i(x).$$

Then, by definition of $\overline{f}$, for each $x \in X$ there are infinitely many n such that $S_n(x) \geq \overline{f}(x) - \varepsilon$. Let

$$N(x) = \min\{n \geq 1 : S_n(x) \geq \overline{f}(x) - \varepsilon\}.$$

Thus $S_{N(x)}(x) \geq \overline{f}(x) - \varepsilon$.

Since $N(x) < \infty$ for all $x \in X$, it follows that there is some $M > 0$ such that

$$\mu(\{x \in X : N(x) > M\}) < \varepsilon.$$

Let

$$B' = B \cup \{x \in X : N(x) > M\},$$

and let

$$S'_n(x) = \frac{1}{n} \sum_{i=0}^{n-1} \mathbb{1}_{B'} \circ T^i(x).$$

Let

$$N'(x) = \begin{cases} N(x) & N(x) \leq M, \\ 1 & N(x) > M. \end{cases}$$

Thus $N'(x) \leq M$ for all $x \in X$. If $N(x) > M$, then $x \in B'$, so

$$S'_{N'(x)}(x) = \mathbb{1}_{B'}(x) = 1 > \overline{f}(x) - \varepsilon.$$

On the other hand, if $N(x) \leq M$, then since $B \subseteq B'$, we have

$$\begin{aligned} S'_{N'(x)}(x) &= \frac{1}{N(x)} \sum_{i=0}^{N(x)-1} \mathbb{1}_{B'} \circ T^i(x) \\ &\geq \frac{1}{N(x)} \sum_{i=0}^{N(x)-1} \mathbb{1}_{B} \circ T^i(x) \\ &\geq \overline{f}(x) - \varepsilon. \end{aligned}$$

Thus $S'_{N'(x)}(x) \geq \overline{f}(x) - \varepsilon$ for all $x \in X$.

Now, define $n_0(x) = 0$ and $n_k(x) = n_{k-1}(x) + N'(T^{n_{k-1}(x)}(x))$ for all $k \geq 1$. Note that the $n_k(x)$'s are the numbers such that $S'_{n_k(x)}(x) \geq \overline{f}(x) - \varepsilon$. Choose n much larger than M, and let $\ell = \max\{k \geq 1 : n_k(x) \leq n\}$.

Since $\overline{f}$ is T-invariant, we have

$$\begin{aligned}
S'_n(x) &= \frac{1}{n}\sum_{i=0}^{n-1} \mathbb{1}_{B'} \circ T^i(x) \\
&\geq \frac{1}{n}\sum_{i=0}^{n_\ell(x)-1} \mathbb{1}_{B'} \circ T^i(x) \\
&= \frac{1}{n}\sum_{i=0}^{\ell-1}\sum_{j=n_i(x)}^{n_{i+1}(x)-1} \mathbb{1}_{B'} \circ T^i(x) \\
&= \frac{1}{n}\sum_{i=0}^{\ell-1} N'(T^{n_i(x)}(x))S'_{N'(T^{n_i(x)}(x))}(T^{n_i(x)}) \\
&\geq \frac{1}{n}\sum_{i=0}^{\ell-1} N'(T^{n_i(x)}(x))(\overline{f}(T^{n_i(x)}(x)) - \varepsilon) \\
&= \frac{1}{n}\sum_{i=0}^{\ell-1}(n_{i+1}(x) - n_i(x))(\overline{f}(x) - \varepsilon) \\
&\geq \frac{n-M}{n}(\overline{f}(x) - \varepsilon).
\end{aligned}$$

Now, because T is measure-preserving, we have

$$\begin{aligned}
\mu(B') &= \frac{1}{n}\int_X \sum_{i=0}^{n-1} \mathbb{1}_{B'} \circ T^i \, d\mu \\
&= \int_X S'_n \, d\mu \\
&\geq \frac{n-M}{n}\left(\int_X \bar{f} \, d\mu - \varepsilon\right).
\end{aligned}$$

Taking the limit as $n \to \infty$, the $\frac{n-M}{n}$ goes to 1, so we have $\mu(B') \geq \int_X \overline{f} \, d\mu - \varepsilon$. But $\mu(B') \leq \mu(B) + \varepsilon$, so $\mu(B) \geq \int_X \overline{f} \, d\mu - 2\varepsilon$. Since ε is arbitrary, we have $\int_X \overline{f} \, d\mu \leq \mu(B)$.

Now we have to show that $\mu(B) \leq \int_X \underline{f} \, d\mu$. To do this, we just repeat the previous argument, replacing B with $B^c = X \setminus B$ and $\mathbb{1}_{B^c}$ in place of $\mathbb{1}_B$. Writing $\overline{g}$ for the analogue of $\overline{f}$ but with $\mathbb{1}_B$ replaced with $\mathbb{1}_{B^c}$, we get $\overline{g}(x) = 1 - \underline{f}(x)$, and

$$1 - \int_X \underline{f}\, d\mu = \int_X \overline{g}\, d\mu \leq \mu(B^c) = 1 - \mu(B),$$

so $\mu(B) \leq \int_X \underline{f}\, d\mu$. This completes the proof. ∎

The next thing to do is to show that if T is ergodic, then f^* is constant almost everywhere, and $f^* = \int_X f\, d\mu$. We have already shown that f^* is T-invariant, so it suffices to learn something about T-invariant functions when T is ergodic:

Proposition 6.11. *Let $(X, \mathcal{A}, \mu, T)$ be a measure-preserving system. Then T is ergodic if and only if every T-invariant measurable function f is constant almost everywhere.*

Proof. Suppose that every T-invariant measurable function f is constant almost everywhere. Let A be a T-invariant set. Since

$$\mathbb{1}_A(T(x)) = \mathbb{1}_{T^{-1}(A)}(x) = \mathbb{1}_A(x),$$

the indicator function $\mathbb{1}_A$ is T-invariant and thus constant almost everywhere. Since $\mathbb{1}_A$ only takes on the values 0 and 1, that means that $\mathbb{1}_A$ is either 0 almost everywhere or 1 almost everywhere. In the former case, $\mu(A) = 0$, and in the latter case, $\mu(A) = 1$. Thus T is ergodic.

Suppose that T is ergodic and f is T-invariant. For each $r \in \mathbb{R}$, let

$$A_r = \{x \in X : f(x) > r\}.$$

Since f is T-invariant, we have $f(x) > r$ if and only if $f(T(x)) > r$, i.e. $x \in A_r$ if and only if $T(x) \in A_r$, i.e. A_r is T-invariant. By ergodicity of T, $\mu(A_r) \in \{0, 1\}$. But if f is not constant almost everywhere, then there would be some $r \in \mathbb{R}$ such that $0 < \mu(A_r) < 1$. ∎

Putting this all together, we see that we have proven the Birkhoff ergodic theorem, at least in the case when f is an indicator function of a measurable subset.

6.5 Problems

(1) Let $X = \{x_1, \ldots, x_r\}$ be a finite set, and let $\sigma : X \to X$ be a permutation. We call σ a *cyclic permutation* if for all i, j, there exists some k such that $\sigma^k(x_i) = x_j$.

(a) Prove that if σ is a cyclic permutation and $\phi : X \to \mathbb{R}$ is any function, then

$$\lim_{n\to\infty} \frac{1}{n} \sum_{j=0}^{n-1} \phi(\sigma^j(x)) = \frac{\phi(x_1) + \cdots + \phi(x_r)}{r}.$$

(b) Prove that if σ is any permutation and $\phi : X \to \mathbb{R}$ is any function, then

$$\lim_{n\to\infty} \frac{1}{n} \sum_{j=0}^{n-1} \phi(\sigma^i(x)) = \frac{\phi(x) + \phi(\sigma(x)) + \cdots + \phi(\sigma^{p-1}(x))}{p},$$

where p is the size of the orbit of x.

(2) (a) Prove that if $T : \mathbb{R}/\mathbb{Z} \to \mathbb{R}/\mathbb{Z}$ is ergodic with respect to Lebesgue measure, then for almost every $x \in \mathbb{R}/\mathbb{Z}$ and every $y \in \mathbb{R}/\mathbb{Z}$ and every $\varepsilon > 0$, there is some positive integer n such that $|T^n(x) - y| < \varepsilon$.

(b) Does there exist an ergodic $T : \mathbb{R}/\mathbb{Z} \to \mathbb{R}/\mathbb{Z}$ such that there exists an $\varepsilon > 0$ such that for all $N > 0$, there exist x, y such that $|T^n(x) - y| \geq \varepsilon$ for all $n \leq N$? Find one or prove that no such T exists.

(3) Let a_n and c_n be sequences, with $\lim_{n\to\infty} c_n = 0$. Let $b_n = a_n + c_n$. Prove that $\liminf_{n\to\infty} a_n = \liminf_{n\to\infty} b_n$ and $\limsup_{n\to\infty} a_n = \limsup_{n\to\infty} b_n$.

(4) Prove that for any finite sequence $d_1, d_2, \ldots, d_k$ of positive integers, there is some positive number b such that for almost all $x \in \mathbb{R}/\mathbb{Z}$, the proportion of windows of length k in the Lüroth series expansion of x equal to $d_1, d_2, \ldots, d_k$ is equal to b. Determine the numbers b, in terms of $d_1, \ldots, d_k$, and prove that your answer is correct.

(5) Prove that irrational rotations on $\mathbb{R}/\mathbb{Z}$ are ergodic. Prove that rational rotations are not ergodic.

(6) From Problem 5, we know that rational rotations are not ergodic. By Proposition 6.11, we know that for any rational number θ, there must be R_θ-invariant functions which are not constant almost everywhere. For each rational rotation, find such an R_θ-invariant function that is not constant almost everywhere.

(7) Characterize the ergodic rotations on $(\mathbb{R}/\mathbb{Z})^n$. If

$$\alpha = (\alpha_1, \ldots, \alpha_n) \in (\mathbb{R}/\mathbb{Z})^n,$$

then the rotation $R_\alpha(x_1, \ldots, x_n)$ is defined to be $(x_1 + \alpha_1 - \lfloor x_1 + \alpha_1 \rfloor, \ldots, x_n + \alpha_n - \lfloor x_n + \alpha_n \rfloor)$.

(8) Prove that there are infinitely many powers of 2 that start with the digits 31415926535.

(9) Prove that $T : [0,1] \to [0,1]$ defined by

$$T(x) = \begin{cases} 2x & 0 \le x < \frac{1}{3}, \\ 2x - \frac{2}{3} & \frac{1}{3} \le x < \frac{1}{2}, \\ 2x - \frac{1}{3} & \frac{1}{2} \le x < \frac{2}{3}, \\ 2x - 1 & \frac{2}{3} \le x \le 1 \end{cases}$$

is ergodic with respect to the Lebesgue measure on $[0,1]$.

(10) Let $T : \mathbb{R}/\mathbb{Z} \to \mathbb{R}/\mathbb{Z}$ be the transformation $T(x) = 2x - \lfloor 2x \rfloor$. Let $\tau : \mathbb{R}/\mathbb{Z} \to \mathbb{N} \cup \{\infty\}$ be defined by $\tau(x) = \min\{n \ge 0 : T^n(x) \in \left(\frac{1}{2}, 1\right)\}$. Prove that $\tau \in L^1(\lambda)$. Show that $2^\tau \notin L^1(\lambda)$. Compute $\int_{\mathbb{R}/\mathbb{Z}} \tau \, d\lambda$.

(11) Let Y be a finite set, and let $X = Y^{\mathbb{N}}$. Prove that every infinite compact subset of X that is invariant under the Bernoulli shift $T : X \to X$ contains a point which is not T-periodic. (A compact subset of X is a finite union of subsets, each of which is a (possibly infinite) intersection of cylinder sets.)

Chapter 7

Invariant Measures

7.1 Continued Fractions

Recall from Chapter 4 the continued fraction expansion of a number $\alpha \in \mathbb{R}/\mathbb{Z}$, which is the sequence $\alpha = [0; a_1, a_2, a_3, \ldots]$, where each a_i is a positive integer. We get it by considering the map $T : \mathbb{R}/\mathbb{Z} \to \mathbb{R}/\mathbb{Z}$ given by

$$T(x) = \frac{1}{x} - \left\lfloor \frac{1}{x} \right\rfloor.$$

Then $a_i = \left\lfloor \frac{1}{T^{i-1}(x)} \right\rfloor$.

We'd like to use our big theorems, especially the ergodic theorem, to discuss properties of the continued fraction expansion. However, there is a problem: T isn't measure-preserving with respect to the Lebesgue measure. For instance, consider the set $I = \left(\frac{1}{2}, 1\right)$. Then

$$T^{-1}(I) = \bigcup_{n=1}^{\infty} \left(\frac{1}{n+1}, \frac{2}{2n+1}\right).$$

Thus we have $\lambda(I) = \frac{1}{2}$, but

$$\lambda(T^{-1}(I)) = \sum_{n=1}^{\infty} \left(\frac{2}{2n+1} - \frac{1}{n+1}\right) = 2\log 2 - 1.$$

Since $\frac{1}{2} \neq 2\log 2 - 1$, T is not measure-preserving.

However, just because T isn't measure-preserving with respect to the Lebesgue measure doesn't mean that all hope is lost: we just need to find another measure for which T is measure-preserving. As we mentioned briefly in Chapter 4, we have such a measure: the Gauß measure. Recall that the Gauß measure is defined by

$$\mu(A) = \frac{1}{\log 2} \int_A \frac{1}{1+x}\, d\lambda,$$

where λ is the Lebesgue measure. We showed in Chapter 4 that T is μ-invariant. So we get to use the ergodic theorem when applied to μ rather than to λ.

The first order of business is to show the following:

Theorem 7.1. *For $\alpha \in \mathbb{R}/\mathbb{Z}$, let $[0; a_1(\alpha), a_2(\alpha), a_3(\alpha), \ldots]$ be its continued fraction expansion. Then for all positive integers k and for almost all $\alpha \in \mathbb{R}/\mathbb{Z}$, we have*

$$\lim_{n\to\infty} \frac{\#\{i \in \{1, 2, \ldots, n\} : a_i(\alpha) = k\}}{n} = \frac{1}{\log 2} \log\left(\frac{(k+1)^2}{k(k+2)}\right).$$

To do this, we first need a lemma that allows us to use the ergodic theorem.

Lemma 7.2. *The continued fraction transformation $T(x) = \frac{1}{x} - \left\lfloor \frac{1}{x} \right\rfloor$ is ergodic with respect to the Gauß measure μ.*

Proof. We already saw in Chapter 4 that for any interval $(a, b) \subseteq [0, 1)$, we have $\mu(T^{-1}((a, b))) = \mu((a - b))$. By the usual results on sufficient semirings, this shows that T is measure-preserving with respect to μ. We now use Knopp's Lemma to prove that T is ergodic with respect to μ; the proof is very similar to the proof that $T(x) = 2x - \lfloor 2x \rfloor$ is ergodic with respect to λ, but we repeat it to review the technique. Let $\mathcal{C}$ be the sufficient semiring of intervals (a, b) of $\mathbb{R}/\mathbb{Z}$ such that, for some integer n, T^n is injective on (a, b) (at least up to a null set), and $T^n((a, b)) = \mathbb{R}/\mathbb{Z}$ (again, up to a null set), for some rational numbers a, b. (If $n = 1$, these are the intervals of the form $\left(\frac{1}{k+1}, \frac{1}{k}\right)$, whereas for $n \geq 2$, these are the intervals (a, b) for which $T^{n-1}((a, b)) = \left(\frac{1}{k+1}, \frac{1}{k}\right)$ for some k.) Let E be such an interval in $\mathcal{C}$, so that $T^n(E) = [0, 1)$ (up to a null set) for some n, n minimal. Suppose that B is a T-invariant set with positive measure. Then we have

$$\frac{\mu(B \cap E)}{\mu(E)} = \frac{\mu(T^{-n}(B) \cap E)}{\mu(E)} = \frac{\mu(B \cap T^n(E))}{\mu(T^n(E))} = \mu(B).$$

This means that $\mu(B \cap E) = \mu(B)\mu(E)$ for every $E \in \mathcal{C}$, so Knopp's Lemma (with $\gamma = \mu(B)$) implies that $\mu(B) = 1$. ■

We now return to the proof of Theorem 7.1.

Proof of Theorem 7.1. Note that we have $a_i(\alpha) = k$ if and only if

$$\frac{1}{k+1} < T^{i-1}(\alpha) \leq \frac{1}{k}.$$

Since T is ergodic with respect to μ, we can apply the ergodic theorem with $f = \mathbb{1}_{\left(\frac{1}{k+1}, \frac{1}{k}\right]}$ to get

$$\begin{aligned}
\lim_{n\to\infty} \frac{\#\{i \in \{1, 2, \dots, n\} : a_i(\alpha) = k\}}{n} &= \lim_{n\to\infty} \sum_{i=0}^{n-1} \mathbb{1}_{\left(\frac{1}{k+1}, \frac{1}{k}\right]}(T^i(\alpha)) \\
&= \int_{[0,1)} \mathbb{1}_{\left(\frac{1}{k+1}, \frac{1}{k}\right]} \, d\mu \\
&= \frac{1}{\log 2} \int_{\frac{1}{k+1}}^{\frac{1}{k}} \frac{dx}{x+1} \\
&= \frac{1}{\log 2} \left[\log\left(1 + \frac{1}{k}\right) \right. \\
&\qquad \left. - \log\left(1 + \frac{1}{k+1}\right) \right] \\
&= \frac{1}{\log 2} \log \frac{(k+1)^2}{k(k+2)},
\end{aligned}$$

as desired. ■

Note that here, we proved the result for μ-almost every $\alpha \in \mathbb{R}/\mathbb{Z}$, rather than for λ-almost every $\alpha \in \mathbb{R}/\mathbb{Z}$. However, this is fine: a measurable subset $A \subseteq \mathbb{R}/\mathbb{Z}$ has Lebesgue measure 0 if and only if it has Gauß measure 0, so in fact the result still hold for λ-almost every $\alpha \in \mathbb{R}/\mathbb{Z}$. Here's a famous closely related theorem regarding the numbers appearing in the continued fraction expansions of almost all numbers in $\mathbb{R}/\mathbb{Z}$.

Theorem 7.3 (Khinchin). *For almost every $\alpha \in \mathbb{R}/\mathbb{Z}$, the limit*

$$\lim_{n\to\infty} \sqrt[n]{\prod_{i=1}^{n} a_i}$$

exists and is equal to

$$\prod_{k=1}^{\infty}\left(\frac{(k+1)^2}{k(k+2)}\right)^{\log_2 k}.$$

Proof. Let $f(x) = \log\left\lfloor \frac{1}{x} \right\rfloor$ in the ergodic theorem. Then for almost all $\alpha \in \mathbb{R}/\mathbb{Z}$, we have

$$\begin{aligned}
\log\left(\lim_{n\to\infty} \sqrt[n]{\prod_{i=1}^{n} a_i(\alpha)}\right) &= \lim_{n\to\infty} \log\left(\sqrt[n]{\prod_{i=1}^{n} a_i(\alpha)}\right) \\
&= \lim_{n\to\infty} \frac{1}{n}\sum_{i=1}^{n} \log(a_i(\alpha)) \\
&= \lim_{n\to\infty} \frac{1}{n}\sum_{i=0}^{n-1} f(T^i(\alpha)) \\
&= \int_{[0,1)} f(x)\,d\mu \\
&= \frac{1}{\log 2}\int_0^1 \frac{f(x)}{x+1}\,dx \\
&= \frac{1}{\log 2}\int_0^1 \frac{\log\left\lfloor\frac{1}{x}\right\rfloor}{x+1}\,dx \\
&= \frac{1}{\log 2}\sum_{k=1}^{\infty}\int_{\frac{1}{k+1}}^{\frac{1}{k}} \frac{\log(k)}{x+1}\,dx \\
&= \frac{1}{\log 2}\sum_{k=1}^{\infty} \log(k)\left[\log\left(1+\frac{1}{k}\right)\right. \\
&\qquad \left. - \log\left(1+\frac{1}{k+1}\right)\right] \\
&= \sum_{k=1}^{\infty} \log\left(\frac{(k+1)^2}{k(k+2)}\right)^{\log_2 k}.
\end{aligned}$$

Now, exponentiating both sides gives the desired result. ■

So, that's the geometric mean of the terms. But what about the arithmetic mean?

Theorem 7.4. *For almost every $\alpha \in \mathbb{R}/\mathbb{Z}$, the sequence*

$$\frac{1}{n}\sum_{i=1}^{n} a_i$$

tends to infinity.

We'll save that for Problem 5.

7.2 Existence of Invariant Measures

When dealing with continued fractions, it's better to work with the Gauß measure rather than the Lebesgue measure, because T is measure-preserving with respect to the Gauß measure but not the Lebesgue measure. We'd like to know whether such measures exist more generally, and if so, how to find them.

Definition 7.5. Let $(X, \mathcal{A})$ be a σ-algebra and $T : X \to X$ a transformation. A measure μ on $(X, \mathcal{A})$ is said to be an *invariant measure* for T if T is measure-preserving with respect to μ, i.e. $\mu(A) = \mu(T^{-1}(A))$ for all $A \in \mathcal{A}$.

Not all transformations have reasonable invariant measures. In the case of transformations on $\mathbb{R}/\mathbb{Z}$, we expect any reasonable invariant measure μ to be a probability measure that is absolutely continuous with respect to the Lebesgue measure, i.e. if $\lambda(A) = 0$, then $\mu(A) = 0$ as well. There are examples of transformations that do not admit an invariant measure of this type. I won't describe one here because the construction is a little bit lengthy, but instead I'll just refer you to Ornstein's paper **[Orn60]** so that you can read the construction on your own. However, if we place some conditions on X, $\mathcal{A}$, and T, then we can guarantee the existence of invariant measures. We'll work with $X = \mathbb{R}/\mathbb{Z}$ and $\mathcal{A}$ the Lebesgue (or Borel) σ-algebra. However, what we do will hold much more generally: when X is a compact metric space. If you are familiar with what that means, then be aware that everything will carry over to that setting. We will also assume that T is a continuous function from X to itself. In order to make this work well for $\mathbb{R}/\mathbb{Z}$, we need to think of it as a circle, rather than a half-open interval, so that 0 is close to 1.

Theorem 7.6 (Krylov–Bogolyubov, [**KB37**]). *If T is a continuous bijection on $\mathbb{R}/\mathbb{Z}$, then there is an invariant measure for T.*

It requires considerable input from real and functional analysis to prove the Krylov–Bogolyubov Theorem rigorously: in particular, we need to know that the space of all Borel probability measures on X is weak*-compact. We will defer that part of the discussion to Appendix A. For now, we will isolate the results we need from Appendix A and use them in our proof.

There is one basic way to construct an invariant object in mathematics: start with something random and average over all the iterates. This will be our technique in the proof.

So, let's start with any probability measure on $\mathbb{R}/\mathbb{Z}$; we'll use the Lebesgue measure λ, but any other measure, even one like δ_0, the point mass at 0, will work just as well. Next, for each positive integer n, we define a measure μ_n by setting

$$\mu_n(A) = \frac{1}{n}\sum_{j=0}^{n-1} \lambda(T^{-j}(A)).$$

Then μ_n is almost invariant because

$$\begin{aligned}\mu_n(T^{-1}(A)) - \mu_n(A) &= \frac{1}{n}\sum_{j=1}^{n} \lambda(T^{-j}(A)) - \frac{1}{n}\sum_{j=0}^{n-1} \lambda(T^{-j}(A)) \\ &= \frac{\lambda(T^{-n}(A)) - \lambda(A)}{n},\end{aligned}$$

which is at most $\frac{1}{n}$ in absolute value. We'd like to take the limit of μ_n and make that our invariant measure, but the problem is that we don't have any guarantee that the limit exists. But that's okay: since $\mu_n(A) \in [0,1]$ for all n, we have a bounded sequence, so by a famous elementary theorem in real analysis (the Bolzano–Weierstraß Theorem), there is an increasing sequence $k_1, k_2, \ldots$ of positive integers such that $\lim_{n\to\infty} \mu_{k_n}(A)$ exists.

We'd like to do the same thing for every measurable subset A, but now we have a new problem: how do we know that the resulting limit function μ is a measure? The issue is that we may have taken completely different sequences for different measurable subsets, so why should they give the compatibility needed to be a measure, e.g. $\mu(A \cup B) = \mu(A) + \mu(B)$ if $A \cap B = \emptyset$? In general, they won't. However, the magic of weak*-compactness is that it is possible to find a sequence that works simultaneously for all measurable subsets A of $\mathbb{R}/\mathbb{Z}$. So, we just use that sequence to take the limit, and that's an invariant measure. Of course, there might be many different sequences that work simultaneously for all $A \in \mathcal{A}$, so then we might end up with several completely different invariant measures.

7.3 Ergodicity of Invariant Measures

Now, suppose that we have a probability measure μ on a σ-algebra $(X, \mathcal{A})$, and suppose we have a transformation $T : X \to X$ which is measure-preserving with respect to μ. As we have seen, T might or might not be ergodic. However, it might be the case that T is measure-preserving with respect to several different probability measures, and it might be ergodic with respect to some of them but not with respect to others. This is because ergodicity means that the only T-invariant measurable subsets of X have measure 0 or 1. However, given a T-invariant subset, it might have measure 0 (say) with respect to some measure but not with respect to a different one.

It turns out that there is a very nice characterization of the ergodic measures. Let us pick a transformation $T : X \to X$ on $(X, \mathcal{A})$, and let $M(X, T)$ denote the set of invariant probability measures for T. The set $M(X, T)$ has certain interesting structure. Suppose we take two measures $\mu_0, \mu_1 \in M(X, T)$. Then for each $t \in [0, 1]$, we can form a new measure $\mu_t \in M(X, T)$ by setting

$$\mu_t(A) = (1 - t)\mu_0(A) + t\mu_1(A).$$

When $t = 0$, we get μ_0, and when $t = 1$, we get μ_1; everything else is somewhere in between.

Let us check that $\mu_t \in M(X, T)$ for all $t \in [0, 1]$. There are many things to verify here: that μ_t is a measure, that it's a probability measure, and that it's invariant for T. These are all straightforward, so let's just

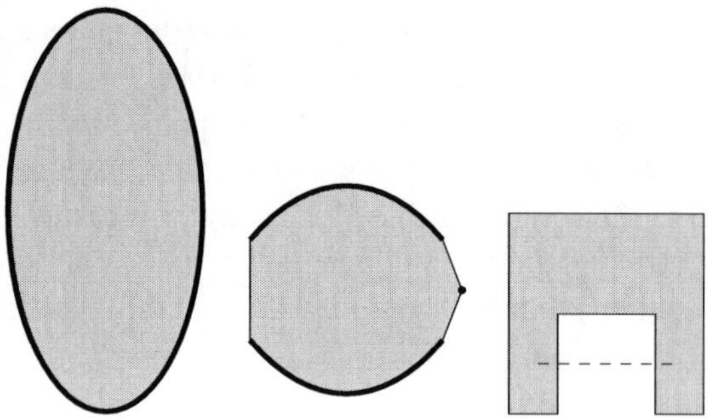

Figure 7.1. Left and Middle: convex sets. Right: a nonconvex set, with a pair of points whose segment is not contained in the set drawn with a dashed line. In both cases, the sets include all the points on the boundary and in the interior. In the convex sets, the extreme points are marked with thick lines, plus the one additional point.

check that it's invariant. Pick some set $A \in \mathcal{A}$. Then we have

$$\begin{aligned} \mu_t(T^{-1}(A)) &= (1-t)\mu_0(T^{-1}(A)) + t\mu_1(T^{-1}(A)) \\ &= (1-t)\mu_0(A) + t\mu_1(A) \\ &= \mu_t(A), \end{aligned}$$

as desired.

A set with this property is called a *convex set*. Not all sets are eligible to be convex sets, since arbitrary sets don't admit a notion of multiplication by real numbers or addition. But measures allow for these operations. Pictorially, we think of a convex set as a geometric object, probably a subset of $\mathbb{R}^n$ or some similar space (possibly an infinite-dimensional one), where convexity means that if we pick any two points in the set, then the line segment connecting them is entirely contained in the set. (See Figure 7.1.)

Now, in a convex set, there are certain special points, known as *extreme points*.

Definition 7.7. Let S be a convex set. An *extreme point* in S is a point $s \in S$ such that if there exist points $s_0, s_1 \in S$ and a real number $t \in (0,1)$ such that $s = (1-t)s_0 + ts_1$, then $s = s_0 = s_1$.

In other words, an extreme point cannot be written as a nontrivial convex combination of two *other* points (i.e. not the point we're starting with) in S.

Example 7.8. Consider the closed disk $\{(x, y) \in \mathbb{R}^2 : x^2 + y^2 \leq 1\}$ in the plane. This is a convex set. Its extreme points are all the points on the boundary circle $x^2 + y^2 = 1$.

Example 7.9. Consider the (filled in) square with vertices $(\pm 1, \pm 1)$. Then the extreme points are the vertices.

Theorem 7.10. *Let T be a continuous function on a compact measure space (think of a circle or closed interval or disk, if you like) X with its Borel σ-algebra. Then the following are equivalent for a measure $\mu \in M(X, T)$:*

(1) *T is ergodic with respect to μ.*

(2) *μ is an extreme point of $M(X, T)$.*

Proof. We first prove that (2) implies (1), or more precisely its contrapositive. Suppose that T is not ergodic. Then there exists a T-invariant measurable subset B of X such that $\mu(T^{-1}(B)) = \mu(B)$ and $0 < \mu(B) < 1$. We'll find two measures μ_0 and μ_1 such that μ is a nontrivial convex combination of μ_0 and μ_1.

To do this, let

$$\mu_0(A) = \frac{\mu(A \cap B)}{\mu(B)}, \qquad \mu_1(A) = \frac{\mu(A \cap B^c)}{\mu(B^c)}.$$

Since $\mu(B)$ and $\mu(B^c)$ are nonzero, these formulae do not involve dividing by zero, and furthermore they are probability measures since $\mu_0(X) = \frac{\mu(X \cap B)}{\mu(B)} = \frac{\mu(B)}{\mu(B)} = 1$ and similarly for $\mu_1(X)$. Furthermore, $\mu_0 \neq \mu_1$ since $\mu_0(B) = 1$ whereas $\mu_1(B) = 0$. In addition, $\mu_0, \mu_1 \in M(X, T)$. For example, μ_0 is T-invariant since for any A, we have

$$\begin{aligned}\mu_0(T^{-1}(A)) &= \frac{\mu(T^{-1}(A)\cap B)}{\mu(B)}\\ &= \frac{\mu(T^{-1}(A)\cap T^{-1}(B))}{\mu(B)}\\ &= \frac{\mu(T^{-1}(A\cap B))}{\mu(B)}\\ &= \frac{\mu(A\cap B)}{\mu(B)}\\ &= \mu_0(A),\end{aligned}$$

as desired; the same calculation shows that μ_1 is also T-invariant.

Finally, we express μ as a convex combination of μ_0 and μ_1: we have

$$\mu(A) = \mu(B)\mu_0(A) + (1-\mu(B))\mu_1(A)$$

since

$$\begin{aligned}\mu(B)\mu_0(A) + (1-\mu(B))\mu_1(A) &= \mu(B)\frac{\mu(A\cap B)}{\mu(B)} + \mu(B^c)\frac{\mu(A\cap B^c)}{\mu(B^c)}\\ &= \mu(A\cap B) + \mu(A\cap B^c)\\ &= \mu(A).\end{aligned}$$

Now, we prove that (1) implies (2). Suppose that T is ergodic with respect to μ, but that μ is a nontrivial convex combination of two other measures in $M(X,T)$, say $\mu = (1-t)\mu_0 + t\mu_1$ where $0 < t < 1$. Let B be any T-invariant set. Since μ is ergodic, we have $\mu(B) \in \{0,1\}$. Thus $\mu_0(B) = \mu_1(B) = \mu(B)$ since the former two are in $[0,1]$ and $0 < t < 1$. Thus both μ_0 and μ_1 must be ergodic.

Now, let A be an arbitrary measurable subset. By the ergodic theorem, we have

$$\mu(A) = \lim_{n\to\infty}\frac{1}{n}\sum_{j=0}^{n-1}\mathbb{1}_A(T^j(x)) \tag{7.1}$$

for almost all $x \in X$. Let's write C for the set of x satisfying (7.1). Then $\mu(C) = 1$, and so $\mu_0(C) = \mu_1(C) = 1$ as well. Again by the ergodic theorem, we have

$$\mu_0(A) = \lim_{n\to\infty}\frac{1}{n}\sum_{j=0}^{n-1}\mathbb{1}_A(T^j(x))$$

for μ_0-almost all $x \in C$. Thus $\mu(A) = \mu_0(A)$. Since A was arbitrary, we have $\mu = \mu_0$, and similarly $\mu = \mu_1$. Thus μ is an extreme point of $M(X,T)$. ■

7.4 Unique Ergodicity

What is the nature of the set $M(X,T)$? Theorem 7.10 gives us a clue, so let's use it to find a pair (X,T) where $M(X,T)$ is fairly large (in particular, not just a single measure).

Consider the rotation R_α on $\mathbb{R}/\mathbb{Z}$. When α is rational, we can easily describe some invariant measures. For concreteness, let's take $\alpha = \frac{1}{2}$. Consider the measure $\mu = \frac{1}{2}\delta_0 + \frac{1}{2}\delta_{\frac{1}{2}}$; that is

$$\mu(A) = \frac{1}{2}\#\left(\left\{0, \frac{1}{2}\right\} \cap A\right).$$

It is easy to see that μ is $R_{\frac{1}{2}}$-invariant, since if A contains 0, then $R_{\frac{1}{2}}^{-1}(A)$ contains $\frac{1}{2}$ and vice versa. Furthermore, $R_{\frac{1}{2}}$ is ergodic with respect to μ, because any $R_{\frac{1}{2}}$-invariant set must either contain both 0 and $\frac{1}{2}$, or else neither of them.

Similarly, for any $\theta \in [0, \frac{1}{2})$, we have a measure μ_θ defined by $\mu_\theta = \frac{1}{2}\delta_\theta + \frac{1}{2}\delta_{\theta+\frac{1}{2}}$. The rotation $R_{\frac{1}{2}}$ is also μ_θ-ergodic.

On the other hand, $R_{\frac{1}{2}}$ is not ergodic with respect to Lebesgue measure, although it is invariant. There are many ways of seeing this; for instance, we can write down an invariant set A with $0 < \lambda(A) < 1$, such as $\left[0, \frac{1}{4}\right) \cup \left[\frac{1}{2}, \frac{3}{4}\right)$. We can also write the Lebesgue measure as a convex combination of two measures. The proof of Theorem 7.10 even tells us how to write it as a convex combination. Let us write $S = \left[0, \frac{1}{4}\right) \cup \left[\frac{1}{2}, \frac{3}{4}\right)$ and $T = S^c = \left[\frac{1}{4}, \frac{1}{2}\right) \cup \left[\frac{3}{4}, 1\right)$. Then let μ_0 and μ_1 be the measures defined by $\mu_0(A) = 2\lambda(A \cap S)$ and $\mu_1(A) = 2\lambda(A \cap T)$. We have $\lambda = \frac{1}{2}\mu_0 + \frac{1}{2}\mu_1$.

I imagine that the μ_θ's are the only ergodic probability measures. An important theorem in functional analysis, the *Krein–Milman Theorem*, would then imply that $M(X,T)$ is the *closed convex hull* (see Figure 7.2) of the μ_θ's, i.e. the smallest closed convex set containing the μ_θ's.

However, sometimes there is just a single invariant measure.

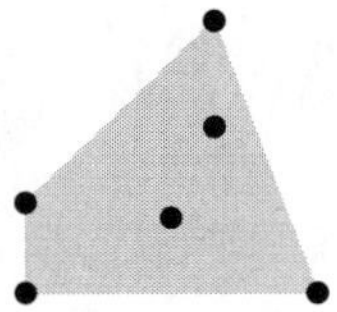

Figure 7.2. The gray region, including its boundary points, is the convex hull of the set of marked points.

Definition 7.11. A transformation T on a measure space $(X, \mathcal{A})$ is said to be *uniquely ergodic* if there is exactly one invariant probability measure for T.

It may seem as though this notion should be called uniquely invariant rather than uniquely ergodic. However, if $M(X, T)$ consists of just a single measure, then it is an extreme point and hence ergodic. By contrast, if $M(X, T)$ consists of more than one point, then it has more than one extreme point and therefore has multiple ergodic measures. So the name is reasonable.

It turns out that irrational rotations on $\mathbb{R}/\mathbb{Z}$ are uniquely ergodic, although this is far from obvious. In order to understand why, we need the following theorem:

Theorem 7.12 (Oxtoby, [**Oxt52**, Theorem 5.1]). *Let X be a compact metric space and $T : X \to X$ a continuous transformation. Then T is uniquely ergodic if and only if for every continuous function $f : X \to \mathbb{R}$, there is a constant $c(f)$, depending on f, such that*

$$\lim_{n\to\infty} \frac{1}{n} \sum_{j=0}^{n-1} f(T^j(x)) = c(f) \tag{7.2}$$

uniformly for all $x \in X$.

Note that this *sounds* like the Birkhoff Ergodic Theorem, which says that the limit should be equal to $\int_X f\, d\mu$, where μ is the unique ergodic measure. However, there's a very important difference here: in the case Birkhoff Ergodic Theorem, convergence only holds for *almost all* $x \in X$, whereas here convergence holds for *all* $x \in X$. As a result, the Birkhoff Ergodic Theorem is useless for proving convergence for any *particular*

x; it only works for almost all of them. If we need to know about convergence for some particular x, we need the stronger hypothesis of unique ergodicity (as well as the other hypotheses of Oxtoby's Theorem).

What we mean by *uniformly* is essentially that the rate of convergence is the same for all $x \in X$. More precisely, uniformity in (7.2) means that

$$\lim_{n\to\infty} \sup_{x\in X} \left| \frac{1}{n} \sum_{j=0}^{n-1} f(T^j(x)) - c(f) \right| = 0.$$

For convergence, we require that for every x, we have

$$\lim_{n\to\infty} \left| \frac{1}{n} \sum_{j=0}^{n-1} f(T^j(x)) - c(f) \right| = 0,$$

so uniformity is stronger because we have to take the supremum before the limit. A typical example of a sequence (of functions) converging non-uniformly to a function is the sequence $f_n(x) = x^n$ on $(-1, 1)$. For every $x \in (-1, 1)$, we have $\lim_{n\to\infty} x^n = 0$. However, this convergence is *not* uniform. For instance, for every n, we can find some value of $x \in (-1, 1)$ such that x^n is as close as desired to 1. Thus we find that in this case,

$$\lim_{n\to\infty} \sup_{x\in(-1,1)} |f_n(x) - 0| = 1$$

and not 0 as we require of uniform convergence. What we get out of uniform convergence is the ability to interchange limits with integrals: if f_n converges to f uniformly on X, then

$$\lim_{n\to\infty} \int_X f_n \, d\mu = \int_X \lim_{n\to\infty} f_n \, d\mu = \int_X f \, d\mu.$$

This is not true in general if convergence isn't uniform, as you will show in Problem 3.

Proof of Theorem 7.12. Let us first assume that (7.2) is true, and we'll show that T is uniquely ergodic. Suppose that we have two invariant probability measures, say μ and ν. Then we have

$$\begin{aligned}\int_X f\,d\mu &= \lim_{n\to\infty} \frac{1}{n} \sum_{j=0}^{n-1} \int_X f \circ T^j\,d\mu \\ &= \int_X \lim_{n\to\infty} \frac{1}{n} \sum_{j=0}^{n-1} f \circ T^j\,d\mu \\ &= \int_X c(f)\,d\mu = c(f),\end{aligned}$$

where $c(f) = \int_X f\,d\mu$. Similarly, $\int_X f\,d\nu = c(f)$. Thus $\int_X f\,d\mu = \int_X f\,d\nu$. Since this holds for all f, it follows that $\mu = \nu$. To see this, note that we can take $f = \mathbb{1}_A$ for any measurable A. Then $\int_X \mathbb{1}_A\,d\mu = \mu(A)$ and $\int_X \mathbb{1}_A\,d\nu = \nu(A)$. So if the two are equal, that means that $\mu(A) = \nu(A)$, and if this holds for all A, then that means that $\mu = \nu$.

The converse requires a bit more input from functional analysis, so we will omit the proof. ■

Let's now use Oxtoby's Theorem to investigate irrational rotations.

Theorem 7.13. *Let $\alpha \in \mathbb{R}/\mathbb{Z}$ be irrational. Then the rotation R_α is uniquely ergodic.*

Proof. We must show that (7.2) holds uniformly in x for all continuous functions $f : \mathbb{R}/\mathbb{Z} \to \mathbb{R}$, i.e.

$$\sup \left| \frac{1}{n} \sum_{j=0}^{n-1} f(T^j(x)) - \int_{\mathbb{R}/\mathbb{Z}} f\,d\mu \right| \to 0 \tag{7.3}$$

as $n \to \infty$. We first prove (7.3) when $f(x) = e^{2\pi ikx}$ for some nonzero integer k. Since $T^j(x) = x + j\alpha \pmod 1$, we have

$$\begin{aligned}\left|\frac{1}{n}\sum_{j=0}^{n-1} f(T^j(x))\right| &= \left|\frac{1}{n}\sum_{j=0}^{n-1} e^{2\pi ik(x+j\alpha)}\right| \\ &= \left|e^{2\pi ikx}\cdot\frac{1}{n}\sum_{j=0}^{n-1} e^{2\pi ik\alpha j}\right| \\ &= \frac{1}{n}\frac{|e^{2\pi ik\alpha n}-1|}{|e^{2\pi ik\alpha}-1|} \\ &\le \frac{1}{n}\frac{2}{|e^{2\pi ik\alpha}-1|}.\end{aligned}$$

Since α is irrational, the denominator is nonzero. Thus the right side tends to 0 as $n \to \infty$. Since $\int_{\mathbb{R}/\mathbb{Z}} f\,d\mu = 0$ (remember the choice of f), we have

$$\sup\left|\frac{1}{n}\sum_{j=0}^{n-1} f(T^j(x)) - \int_{\mathbb{R}/\mathbb{Z}} f\,d\mu\right| \le \frac{1}{n}\frac{2}{|e^{2\pi ik\alpha}-1|} \to 0$$

as $n \to \infty$. Thus the result holds for this choice of function f. It also holds when f is a constant function, i.e. when $k = 0$. So it holds for all integers k. Similarly, it holds for finite linear combinations of exponential functions $e^{2\pi ikx}$, also known as *trigonometric polynomials.*

To complete the proof for an arbitrary function, we make use of another important theorem from analysis, the Stone–Weierstraß Theorem [**Pug15**, Theorem 20, pp. 234–238]. The Stone–Weierstraß Theorem says that every continuous function on $\mathbb{R}/\mathbb{Z}$ is a uniform limit of these trigonometric polynomials. ■

7.5 Problems

(1) Let $(X, \mathcal{A})$ be a σ-algebra, and let $x \in X$. Let T be a transformation on $(X, \mathcal{A})$. When is the point mass δ_x an invariant measure for T? More generally, if $x_1, \ldots, x_n \in X$, when is the sum $\frac{1}{n}(\delta_{x_1} + \cdots + \delta_{x_n})$ an invariant measure for T?

(2) Prove that the Lebesgue measure is the only measure on $\mathbb{R}/\mathbb{Z}$ under which all rational rotations are invariant. This is a sort of complement to Theorem 7.13.

(3) (a) Find a sequence of functions $f_n : (-1,1) \to \mathbb{R}_{\geq 0}$ such that for all $x \in (-1,1)$, $\lim_{n\to\infty} f_n(x)$ exists, but

$$\lim_{n\to\infty} \int_{(-1,1)} f_n \, d\lambda \neq \int_{(-1,1)} \lim_{n\to\infty} f_n \, d\lambda.$$

(Recall the Dominated Convergence Theorem when constructing your example.)

(b) Prove that if $f_n \to f$ uniformly, then

$$\lim_{n\to\infty} \int_{(-1,1)} f_n \, d\lambda = \int_{(-1,1)} \lim_{n\to\infty} f_n \, d\lambda.$$

(4) Let S be the set of $\alpha \in \mathbb{R}/\mathbb{Z}$ whose continued fraction coefficients a_i are bounded. Prove that $\lambda(S) = 0$.

(5) Prove Theorem 7.4.

(6) Give an example of a transformation $T : X \to X$ on a compact metric space such that $\frac{1}{n}\sum_{j=0}^{n-1} \phi(T^j(x))$ converges uniformly for every continuous function $\phi : X \to \mathbb{R}$, but where T is not uniquely ergodic. (Note that by Oxtoby's Theorem, this means that the sum cannot converge uniformly to a constant function.) Can you do it when $X = \mathbb{R}/\mathbb{Z}$?

(7) Let d be an integer from 1 to 9. What is the proportion of powers of 2 beginning with the digit d? Prove that your answer is correct. (Hint: Use Oxtoby's Theorem.)

(8) Recall the induced transformation from Chapter 5, defined as follows: suppose $(X, \mathcal{A}, \mu, T)$ is a measure-preserving system. Let $A \in \mathcal{A}$ have $\mu(A) > 0$. Then for almost all $x \in A$, there is some positive integer $n_A(x)$ such that $T^{n_A(x)}(x) \in A$; let $n_A(x)$ be the smallest such positive integer. We define $T_A(x) = T^{n_A(x)}(x)$, which is defined for almost all $a \in A$. In this problem, we'll do a sort of converse.

(a) Suppose that $T : X \to X$ is a measurable (but not necessarily measure-preserving) transformation on X, with the property that for almost all $x \in A$, there is a positive integer $n(x)$ such that $T^{n(x)} \in A$, and let $n(x)$ be any such positive integer subject to the constraint that $n : A \to \mathbb{N}$ is a measurable function. Let $A_n = \{x \in A : n(x) = n\}$; this $A_n \in \mathcal{A}$. Define $T_A : A \to A$ by $T_A(x) = T^{n(x)}(x)$. Suppose that $T_A : A \to A$ is measure-preserving on A with respect to μ (or, more precisely, its restriction to A). Define a function ν_A on X by

$$\nu_A(B) = \sum_{n=0}^{\infty} \sum_{k>n} \mu(T^{-n}(B) \cap A_k).$$

Prove that ν_A is a measure on X.

(b) Prove that ν_A is T-invariant.

(c) Prove that if $n(x)$ is the first return time (i.e. $n(x)$ is chosen to be minimal), then $\mu(B) = \nu_A(B)$ for every measurable $B \subseteq A$.

(d) Prove that if μ is the restriction of a measure ν on X to A, then $\nu_A(B) \leq \nu(B)$ for every $B \in \mathcal{A}$. (That is, B is only a subset of X, not necessarily a subset of A.)

(9) Let $\beta > 1$ be a real number. Consider the transformation T_β on $\mathbb{R}/\mathbb{Z}$, defined by $T_\beta(x) = \beta x - \lfloor \beta x \rfloor$.

(a) Let $\beta = \frac{1+\sqrt{5}}{2}$, and define a function $g_\beta : \mathbb{R}/\mathbb{Z} \to \mathbb{R}/\mathbb{Z}$ by

$$g_\beta(x) = \begin{cases} \frac{5+3\sqrt{5}}{10} & 0 \leq x \leq \frac{\sqrt{5}-1}{2}, \\ \frac{5+\sqrt{5}}{10} & \frac{\sqrt{5}-1}{2} < x < 1. \end{cases}$$

Prove that the function μ_β on the Lebesgue σ-algebra $\mathcal{A}$ defined by

$$\mu_\beta(A) = \int_A g_\beta \, d\lambda$$

is a probability measure on $\mathbb{R}/\mathbb{Z}$.

(b) Prove that T_β is invariant with respect to μ_β.

(c) Find an invariant measure for T_β for other values of β. (Hint: come up with a piecewise linear function similar to g_β.)

(10) Let $\beta = \frac{1+\sqrt{5}}{2}$ again. Let

$$X = \left(\left[0, \frac{1}{\beta}\right) \times [0,1)\right) \cup \left(\left[\frac{1}{\beta}, 1\right) \times \left[0, \frac{1}{\beta}\right)\right).$$

Put the Lebesgue σ-algebra and measure $\lambda \times \lambda$ on X. Consider the transformation $T : X \to X$ defined by

$$T(x,y) = \begin{cases} \left(\beta x, \frac{y}{\beta}\right) & (x,y) \in \left[0, \frac{1}{\beta}\right) \times [0,1), \\ \left(\beta x - 1, \frac{1+y}{\beta}\right) & (x,y) \in \left[\frac{1}{\beta}, 1\right) \times \left[0, \frac{1}{\beta}\right). \end{cases}$$

(a) Prove that T is measure-preserving with respect to $\lambda \times \lambda$.
(b) Write down an explicit formula for the induced transformation on $[0,1) \times \left[0, \frac{1}{\beta}\right)$.

(11) Again, let $\beta = \frac{1+\sqrt{5}}{2}$, and let T_β be as in Problem 9. For $x \in [0,1)$, define a sequence of numbers $b_1(x), b_2(x), \ldots \in \{0,1\}$ by

$$b_k(x) = \begin{cases} 0 & 0 \le T_\beta^{k-1}(x) < \frac{1}{\beta}, \\ 1 & \frac{1}{\beta} \le T_\beta^{k-1}(x) < 1. \end{cases}$$

For each positive integer j, prove that for almost all $x \in \mathbb{R}/\mathbb{Z}$,

$$\lim_{n\to\infty} \frac{1}{n} \sum_{i=0}^{n-1} \prod_{k=0}^{j-1} (1 - b_{i+k}(x)) \tag{7.4}$$

exists and is constant. (In other words, prove that the proportion of i's are $b_i, b_{i+1}, \ldots, b_{i+j-1}$ all equal to 0 is the same for almost all x.) Determine what this constant is, in terms of j.

(12) In this problem, we'll generalize the result of Problem 11. Let $\beta > 1$ be irrational. For $x \in [0,1)$, define $b_k(x)$ to be $b_k(x) = \lfloor \beta T_\beta^{k-1}(x) \rfloor$, where T_β is as in Problem 9.

(a) Prove that $x = \sum_{k=1}^{\infty} \frac{b_k(x)}{\beta^k}$. Thus we think of the b_k's as giving the *base-β expansion* (or simply β-expansion) of x.
(b) In terms of β, what are the allowable patterns of consecutive digits in the β-expansion? For example, can 3562 appear consecutively in the β-expansion of some $x \in [0,1)$?
(c) Prove that the limit in (7.4) exists for any β-expansion and is constant for almost all $x \in [0,1)$. Evaluate the limit.

Chapter 8

Mixing

8.1 Mixing

Recall that a measure-preserving system $(X, \mathcal{A}, \mu, T)$ is ergodic if and only if for all measurable subsets $A, B \subseteq X$, we have

$$\lim_{n\to\infty} \frac{1}{n} \sum_{j=0}^{n-1} \mu(T^{-j}(A) \cap B) = \mu(A)\mu(B).$$

In general, we cannot replace the left side by $\lim_{n\to\infty} \mu(T^{-n}(A) \cap B)$; we have to take an average. Recall, for instance, that if $T = R_\alpha$ is an irrational rotation and A is a small interval in $\mathbb{R}/\mathbb{Z}$, then $T^{-n}(A) \cap A = \varnothing$ for infinitely many values of n, so in particular its measure does not tend to $\lambda(A)^2$.

However, there are many cases in which we can replace the average with the single term $\mu(T^{-n}(A) \cap B)$. This is the idea of mixing.

Definition 8.1. A measure-preserving system $(X, \mathcal{A}, \mu, T)$ is said to be *mixing* if, for all $A, B \in \mathcal{A}$, we have

$$\lim_{n\to\infty} \mu(T^{-n}(A) \cap B) = \mu(A)\mu(B).$$

If $\mu(B) > 0$ and we divide by $\mu(B)$, the condition for mixing becomes

$$\lim_{n\to\infty} \frac{\mu(T^{-n}(A) \cap B)}{\mu(B)} = \mu(A).$$

This says that the proportion of B that is in A after many iterates is roughly the measure of A. We can interpret this as saying that we've forgotten that we started from B, and we're now just at a random point in X, with probability based on the measure μ.

We can also interpret this in terms of probability. If we have two events A and B, then they are independent if knowing that A happens does not give us further information about B, i.e.

$$\mathbb{P}(B \mid A) = \mathbb{P}(B).$$

Since $\mathbb{P}(B \mid A) = \frac{\mathbb{P}(A \cap B)}{\mathbb{P}(A)}$, this just says that

$$\frac{\mathbb{P}(A \cap B)}{\mathbb{P}(A)} = \mathbb{P}(B),$$

or

$$\mathbb{P}(A \cap B) = \mathbb{P}(A)\mathbb{P}(B).$$

So mixing is an asymptotic version of independence: after any fixed number n of iterates of T, $T^{-n}(A)$ and B might not be independent, but they get closer and closer to independence as n tends to ∞, so in the limit they are independent.

Note that several familiar examples of transformations are mixing. For example, the decimal expansion transformation $T(x) = 10x - \lfloor 10x \rfloor$ and the continued fraction transformation $T(x) = \frac{1}{x} - \left\lfloor \frac{1}{x} \right\rfloor$ (with respect to the Gauß measure) are both mixing. But, as we mentioned above, irrational rotations are ergodic but not mixing.

8.2 Types of Convergence

Before we study mixing in more detail and also introduce the related notion of weak mixing, let's talk about various types of convergence of sequences. Let's suppose we have a sequence $a_0, a_1, a_2, \ldots$ of real numbers. The standard way that it can converge to some real number a is that we have

$$\lim_{i \to \infty} a_i = a,$$

or, to make it look more like the other types,

$$\lim_{i \to \infty} (a_i - a) = 0.$$

Another notion of convergence is *strong Cesàro convergence*. We say that a_i is *strongly Cesàro convergent* to a if

$$\lim_{n\to\infty} \frac{1}{n} \sum_{i=0}^{n-1} |a_i - a| = 0.$$

Then there's also *weak Cesàro convergence*, which means that

$$\lim_{n\to\infty} \frac{1}{n} \sum_{i=0}^{n-1} (a_i - a) = 0.$$

If a_i converges to a, then it strongly Cesàro converges to a, and if a_i strongly Cesàro converges to a, it also weakly Cesàro converges to a. However, these implications only work in one way. Let's have a look:

Example 8.2. Let $a_0, a_1, a_2, \dots$ be the sequence defined by

$$a_i = \begin{cases} 1 & i \text{ is a power of } 2, \\ 0 & \text{otherwise.} \end{cases}$$

Then a_i does not converge to 0 in the usual sense, but it is strongly Cesàro convergent to 0. Let's check this using the definition. Let 2^r be the largest power of 2 that is at most $n-1$. Then we have

$$\frac{1}{n} \sum_{i=0}^{n-1} |a_i - 0| = \frac{1}{n}(r+1) = \frac{\lceil \log_2(n) \rceil}{n} \to 0$$

as $n \to \infty$.

Example 8.3. Let $a_i = (-1)^i$. Then a_i is weakly Cesàro convergent to 0, but not strongly Cesàro convergent. Let's check first that it's weakly Cesàro convergent. We have

$$\frac{1}{n} \sum_{i=0}^{n-1} (a_i - 0) = \frac{1}{n} \sum_{i=0}^{n-1} (-1)^i \le \frac{1}{n}.$$

(It's either 0 or $\frac{1}{n}$, depending on the parity of n.) Thus it tends to 0 as $n \to \infty$. On the other hand, if we include absolute values, we have

$$\frac{1}{n} \sum_{i=0}^{n-1} |a_i - 0| = \frac{1}{n} \sum_{i=0}^{n-1} 1 = 1.$$

One way of interpreting strong Cesàro convergence is that it means that "most" of the a_i's tend to a, but there can be rare exceptions. That's what we see in our example of a sequence that is strongly Cesàro convergent but not convergent: if we remove the powers of 2 (which are rare in the nonnegative integers), then we get a convergent sequence.

Definition 8.4. Let A be a set of nonnegative integers. We say that the *density* $\delta(A)$ of A is

$$\delta(A) = \lim_{n\to\infty} \frac{\#A \cap \{0, 1, \dots, n-1\}}{n},$$

if this limit exists.

Alternatively, we can write

$$\delta(A) = \lim_{n\to\infty} \frac{1}{n} \sum_{i=0}^{n-1} \mathbb{1}_A(i).$$

This is a useful characterization, as we shall see.

For instance, the primes, perfect squares, and powers of 2 all have density 0. Some sets of nonnegative integers don't have a density in this sense; a typical example of such a set is the numbers beginning with a 1. (There is another sense of density, called *logarithmic density*, where this set does have a well-defined density of $\log_{10}(2)$, but we don't need that here.)

We can use density to give an equivalent formulation of Cesàro convergence in a certain case:

Definition 8.5. A sequence $a_0, a_1, a_2, \dots$ of real numbers is said to *converge in density* to a real number a if there exists a subset D of $\mathbb{N}$ with $\delta(D) = 0$ such that for all $\varepsilon > 0$, there is an N such that whenever $n > N$ and $n \notin D$, we have $|a_n - a| < \varepsilon$. We write

$$\lim_{\substack{n\to\infty \\ n\notin D}} a_n = a$$

in this case. When we want to emphasize the choice of D, we say that $a_0, a_1, a_2, \dots$ converges to a outside D.

Proposition 8.6. *A bounded sequence of nonnegative real numbers Cesàro converges to 0 if and only if it converges in density to 0.*

Note that we need not distinguish between strong and weak Cesàro convergence here, because for sequences of nonnegative real numbers converging to 0, the two notions are the same: the absolute value is irrelevant since all the terms in the sum are already nonnegative.

Proof. Suppose that $a_0, a_1, a_2, \dots$ is a bounded sequence, say with $a_n \le b$ for all n. Furthermore, let's suppose that it converges in density to 0, with D being the density-0 set we have to ignore. Then we have

$$\begin{aligned}
\lim_{n\to\infty} \frac{1}{n} \sum_{i=0}^{n-1} a_i &= \lim_{n\to\infty} \frac{1}{n} \left[\sum_{\substack{0\le i\le n-1 \\ i\notin D}} a_i + \sum_{\substack{0\le i\le n-1 \\ i\in D}} a_i \right] \\
&= \lim_{n\to\infty} \frac{1}{n} \sum_{\substack{0\le i\le n-1 \\ i\notin D}} a_i + \lim_{n\to\infty} \frac{1}{n} \sum_{\substack{0\le i\le n-1 \\ i\in D}} a_i \\
&\le \lim_{n\to\infty} \frac{1}{n} \sum_{\substack{0\le i\le n-1 \\ i\notin D}} a_i + \lim_{n\to\infty} \frac{1}{n} \sum_{\substack{0\le i\le n-1 \\ i\in D}} b \\
&\le 0 + 0 \\
&= 0.
\end{aligned}$$

The limit is also ≥ 0 because all the terms are nonnegative.

Now, for the converse direction, suppose that

$$\lim_{n\to\infty} \frac{1}{n} \sum_{i=0}^{n-1} a_i = 0.$$

We must construct a zero density set D. For each positive integer k, define

$$D_k = \left\{ i : a_i > \frac{1}{k} \right\}.$$

Note that $\frac{1}{k}\mathbb{1}_{D_k}(i) < a_i$. Thus we have

$$\frac{1}{k} \lim_{n\to\infty} \frac{1}{n} \sum_{i=0}^{n-1} \mathbb{1}_{D_k}(i) = \lim_{n\to\infty} \sum_{i=0}^{n-1} \frac{1}{k} \mathbb{1}_{D_k}(i) \le \lim_{n\to\infty} \frac{1}{n} \sum_{i=0}^{n-1} a_i = 0,$$

so each D_k has density 0. It follows that for each k, we can find an integer N_k with $N_k > N_{k-1} > 0$ (and $N_0 = 0$) so that

$$\frac{1}{n}\sum_{i=0}^{n-1} \mathbb{1}_{D_k}(i) < \frac{1}{k}$$

for all $n > N_k$. Now let

$$D = \bigsqcup_{k>0} D_k \cap [N_k, N_{k+1}).$$

Then a_i converges to 0 outside D, and D has density 0, as you will verify in Problem 7. ■

8.3 Weak Mixing

Weak mixing is a notion between ergodicity and mixing: if a transformation is mixing, then it is weakly mixing, and if a transformation is weakly mixing, then it is ergodic. The definition of weak mixing is expressed in terms of Cesàro convergence. Consider the sequence $a_i = \mu(T^{-i}(A) \cap B)$. On average, a_i is equal to $\mu(A)\mu(B)$ if T is ergodic. But in what sense does a_i converge to $\mu(A)\mu(B)$? There are several possibilities:

- $\lim_{n\to\infty} a_n = \mu(A)\mu(B)$. If this happens for all $A, B \in \mathcal{A}$, that means that T is mixing, for this is simply the definition of mixing.

- a_n strongly Cesàro converges to $\mu(A)\mu(B)$, i.e.

  $$\lim_{n\to\infty} \frac{1}{n}\sum_{i=0}^{n-1} |\mu(T^{-i}(A) \cap B) - \mu(A)\mu(B)| = 0.$$

  If this happens for all $A, B \in \mathcal{A}$, we say that T is *weakly mixing*.

- a_n weakly Cesàro converges to $\mu(A)\mu(B)$, i.e.

  $$\lim_{n\to\infty} \frac{1}{n}\sum_{i=0}^{n-1} \mu(T^{-i}(A) \cap B) = \mu(A)\mu(B).$$

  If this happens for all $A, B \in \mathcal{A}$, this is the same as saying that T is ergodic.

Thus we have the implications

$$\text{mixing} \quad \Rightarrow \quad \text{weak mixing} \quad \Rightarrow \quad \text{ergodic.}$$

An interesting question is to determine whether the reverse directions hold: is a weakly mixing transformation necessarily mixing? Is an ergodic transformation necessarily weakly mixing? The answers to both of these questions is no. We know that the answer to the corresponding sequence versions are no: a strongly Cesàro convergent sequence is not necessarily convergent, and a weakly Cesàro convergent sequence is not necessarily strongly Cesàro convergent. But that doesn't say anything about the transformation version. To do that, we have to construct the relevant transformations.

Our next task will be to construct an ergodic transformation that is not weakly mixing. A good candidate is an irrational rotation R_α on $\mathbb{R}/\mathbb{Z}$, because we know that this transformation is ergodic but not mixing. So we have two possibilities: either it's an ergodic transformation that isn't weakly mixing, or else it's a weakly mixing transformation that isn't mixing. It turns out that the former of these is correct: it's not weakly mixing. But intuitively, you should already believe that irrational rotations aren't weakly mixing, because of the interpretation of strongly Cesàro convergence being that most of the time it's convergent, just that there are rare exceptions. This doesn't seem so likely in the case of irrational rotations, because for small intervals A and B, it seems like $R_\alpha^{-i}(A) \cap B$ will be empty for many values of i. However, a proof requires a bit of care. It will be helpful to have some preliminary results.

Proposition 8.7. *Let $(X, \mathcal{A}, \mu, T)$ be a measure-preserving system. Then the following are equivalent:*

(1) *T is weakly mixing.*

(2) *For every pair of subsets $A, B \in \mathcal{A}$, there is a zero-density set D, depending on A and B, such that*

$$\lim_{\substack{i \to \infty \\ i \notin D}} \mu(T^{-i}(A) \cap B) = \mu(A)\mu(B).$$

(3) *For every pair of subsets $A, B \in \mathcal{A}$, we have*

$$\lim_{n\to\infty} \frac{1}{n} \sum_{i=0}^{n-1} \left(\mu(T^{-i}(A) \cap B) - \mu(A)\mu(B)\right)^2 = 0.$$

Proof. The equivalence of (1) and (2) follows from Proposition 8.6, applied to the sequence

$$a_i = |\mu(T^{-i}(A) \cap B) - \mu(A)\mu(B)|,$$

which is bounded by 1. So we must show that (1) (or (2)) is equivalent to (3). Suppose that (1) holds, i.e. that T is weakly mixing. We also get to assume that (2) holds, so take the D in (2). We have

$$\lim_{\substack{i\to\infty \\ i\notin D}} (\mu(T^{-i}(A) \cap B) - \mu(A)\mu(B)) = 0,$$

so we may freely square this to get

$$\lim_{\substack{i\to\infty \\ i\notin D}} (\mu(T^{-i}(A) \cap B) - \mu(A)\mu(B))^2 = 0.$$

By Proposition 8.6 again, this time applied to the sequence

$$b_i = (\mu(T^{-i}(A) \cap B) - \mu(A)\mu(B))^2,$$

we know that the sequence b_i Cesàro converges to 0, which is exactly the statement of (3). The converse (3) implies (1) or (2) is simply the reverse of this argument. ∎

Here is another notion equivalent to weak mixing, known as *double ergodicity*.

Definition 8.8. A measure-preserving system $(X, \mathcal{A}, \mu, T)$ is said to be *doubly ergodic* if, for all measurable subsets A and B with positive measure, there is a positive integer n such that

$$\mu(T^{-n}(A) \cap A) > 0, \qquad \mu(T^{-n}(A) \cap B) > 0. \tag{8.1}$$

Note that the times n have to be the same in (8.1), so that both A and B can reach A in the same amount of time.

Theorem 8.9. *If T is weakly mixing, then it is doubly ergodic.*

Remark 8.10. The converse is also true with some additional hypotheses: if T is an invertible doubly ergodic transformation on $(\mathbb{R}/\mathbb{Z}, \mathcal{L}, \mu)$, then it is weakly mixing. See [**Sil08**, Theorem 6.4.2, pp. 216–218] and the references therein for a proof.

Proof. Suppose that T is weakly mixing. Take two measurable subsets A and B with $\mu(A), \mu(B) > 0$, and let $D(A, A)$ and $D(A, B)$ be as in part (2) of Proposition 8.7 for the subsets A and A, and A and B, respectively, so that we have

$$\lim_{\substack{i\to\infty \\ i\notin D(A,A)}} \mu(T^{-i}(A) \cap A) = \mu(A)^2,$$

$$\lim_{\substack{i\to\infty \\ i\notin D(A,B)}} \mu(T^{-i}(A) \cap B) = \mu(A)\mu(B).$$

Because $D(A, A) \cup D(A, B)$ still has density 0, there is some $n \notin D(A, A) \cup D(A, B)$ such that $\mu(T^{-n}(A) \cap A) > 0$ and $\mu(T^{-n}(A) \cap B) > 0$. Thus T is doubly ergodic. ∎

Finally, we can show that irrational rotations are not weakly mixing.

Theorem 8.11. *Irrational rotations are not doubly ergodic, and thus not weakly mixing.*

Proof. Let α be irrational. Let $A = \left[0, \frac{1}{8}\right)$ and $B = \left[\frac{1}{2}, \frac{5}{8}\right)$. Then for all n, $R_\alpha^{-n}(A)$ is an interval of length $\frac{1}{8}$, so it cannot intersect both A and B. Thus R_α is not doubly ergodic and thus not weakly mixing. ∎

8.4 Weakly Mixing Does Not Imply Mixing

We have seen that ergodicity does not imply weak mixing. It is also true that weak mixing does not imply mixing, and we now give an example of a transformation that is weakly mixing but not mixing. This is a more challenging task than giving a transformation that is ergodic but not weakly mixing, but we can do it nonetheless. This is an interesting challenge, because it requires that, for *most but not all* large values of n, $\mu(T^{-n}(A) \cap B)$ should be close to $\mu(A)\mu(B)$.

To do this, we introduce two measure-preserving transformations on $[0, 1)$. The first one, the *dyadic odometer*, is a warmup for the second

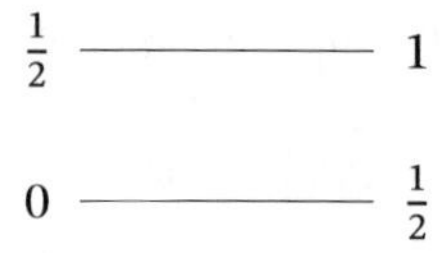

Figure 8.1. A pictorial interpretation of the partial transformation $T(x)$ defined by $T(x) = x + \frac{1}{2}$ if $0 \leq x < \frac{1}{2}$ and undefined otherwise

one, the *Chacón transformation.* To define them, it will be helpful to view these transformations pictorially. What we will do is to draw certain subintervals of $[0, 1)$ stacked on top of each other. What the transformation does is to send a point in one of these subintervals to the point directly above it. The transformation is not defined at a point on the top level, but this will turn out not to be a problem.

As an example, let's divide the interval $[0, 1)$ into subintervals $I_1 = \left[0, \frac{1}{2}\right)$ and $I_2 = \left[\frac{1}{2}, 1\right)$ and put I_1 below I_2, as shown in Figure 8.1. The implied transformation sends x to $x + \frac{1}{2}$ if $0 \leq x < \frac{1}{2}$ and is undefined otherwise. In what follows, all the intervals in a picture will always be the same length, so we send one interval to the other by means of a measure-preserving linear function.

This is the first step of the dyadic odometer. Let's call this partially defined transformation $T_1(x)$. To create T_2, which is also only partially defined, cut the diagram in half from left to right. Then we end up with two stacks. The left stack has $\left[0, \frac{1}{4}\right)$ on the bottom and $\left[\frac{1}{2}, \frac{3}{4}\right)$ on the top, and the right stack has $\left[\frac{1}{4}, \frac{1}{2}\right)$ on the bottom and $\left[\frac{3}{4}, 1\right)$ on the top. Now, place the right stack on top of the left stack, as shown in Figure 8.2.

We can iterate this process. To get from T_n to T_{n+1}, take the stack of intervals in T_n, cut it in half from left to right, then place the right stack on top of the left stack. It's clear that $T_n(x) = T_{n+1}(x)$ wherever both of them are defined, since if one point is directly above another point, this iterative process will never change that. All the iterations do is to allow T_{n+1} to be defined at more points than T_n is. Furthermore, for every $x \in [0, 1)$, $T_n(x)$ is defined for some n: T_n is defined everywhere except on the top interval, which is $[1 - 2^{-n}, 1)$, and no point in $[0, 1)$ is in $\bigcap_{n=1}^{\infty} [1 - 2^{-n}, 1)$: that intersection is empty. So we can define the

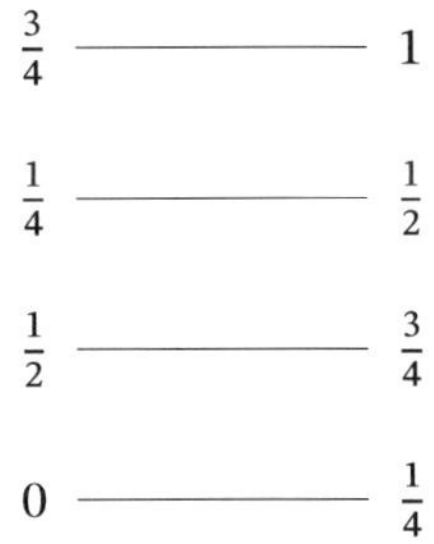

Figure 8.2. The transformation T_2

full dyadic odometer to be the transformation $T(x) = \lim_{n\to\infty} T_n(x)$, or simply $T_n(x)$ where n is chosen so that $T_n(x)$ is defined.

It's pretty clear intuitively that the dyadic odometer is measure-preserving, because $T^{-1}(A)$ just pushes A down a level. The only thing to watch out for is that we may have points on the bottom level, so that if x is on the bottom level then $T_n^{-1}(x)$ is empty. However, not many points will stay at the bottom forever: eventually they'll be on the right side of the bottom interval and then they'll get moved. The only exception is 0, which always stays at the very bottom of the stack. However, it's perfectly okay in a measure-preserving transformation for the preimage of a single point to be empty, since the (Lebesgue) measure of a point is 0.

Now let's describe the Chacón transformation, which is constructed in a similar manner. To do this, we start by putting the first $\frac{2}{3}$ of the interval (i.e. $\left[0, \frac{2}{3}\right)$ on the bottom of a stack, and that's all; we reserve $\left[\frac{2}{3}, 1\right)$ for later. The zeroth-level transformation T_0 does nothing.

To construct T_1, we divide the stack we already have, i.e. $\left[0, \frac{2}{3}\right)$, into three equal subintervals, namely $\left[0, \frac{2}{9}\right)$, $\left[\frac{2}{9}, \frac{4}{9}\right)$, and $\left[\frac{4}{9}, \frac{2}{3}\right)$, and then place an equal-length interval on top of the middle piece, so that's $\left[\frac{2}{3}, \frac{8}{9}\right)$. Now we have three substacks, where the left and right substacks have height 1 and the middle one has height 2. Now we put the left stack on the bottom, then the middle stack, then finally the right stack on the top; see Figure 8.3.

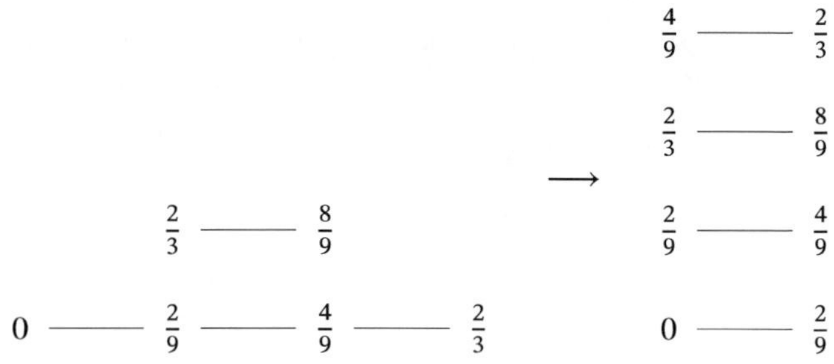

Figure 8.3. Stage 1 of constructing the Chacón transformation

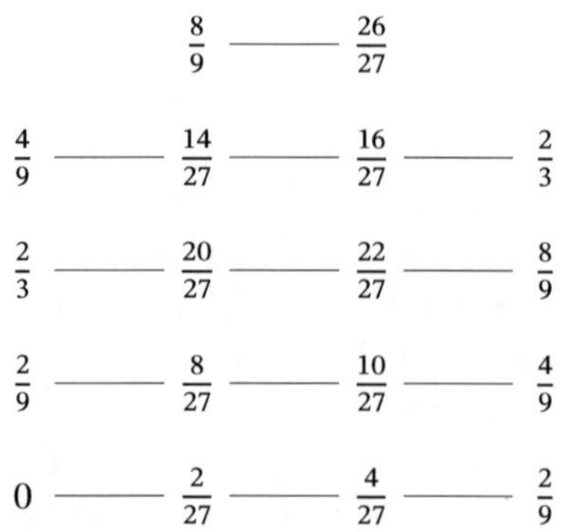

Figure 8.4. Stage 2 of constructing the Chacón transformation

Now we simply iterate the procedure. To construct T_{n+1} from T_n, we divide our stack into three equal pieces from left to right. We then put a new interval, from the left of the remaining piece $[1 - 3^{-(n+1)}, 1)$ on top of the middle stack, so that's $[1 - 3^{-(n+1)}, 1 - 3^{-(n+2)})$. Then we put the left stack on the bottom, followed by the middle stack, and finally the right stack on the top. See Figure 8.4 for stage 2 of this construction.

Just as with the dyadic odometer, if $T_n(x)$ is defined, then $T_n(x) = T_{n+1}(x)$, but T_{n+1} is defined at more points. Furthermore, for every $x \in [0, 1)$, $T_n(x)$ is defined for some n, so we have a transformation T defined by $T(x) = \lim_{n\to\infty} T_n(x)$ defined on all of $[0, 1)$. This is the Chacón transformation. It's measure-preserving for all the same reasons that the dyadic odometer is: $T^{-1}(A)$ just means going down a level in the stack.

The Chacón transformation is an example of a transformation that is weakly mixing but not mixing. In order to prove these facts, we need a preliminary lemma. The following notation will also be helpful: we call the intervals in the nth iteration of the construction the *levels*, and we write h_n for the height of the stack at the nth iteration. Concretely, the intervals at the nth level are all of width $\frac{2}{3^{n+1}}$, and the height h_n is $\sum_{k=0}^{n} 3^k = \frac{3^{n+1}-1}{2}$.

Lemma 8.12. *Let I and J be levels of the nth iterate of the Chacón transformation, so in particular they are intervals of width $\lambda(I) = \lambda(J) = \frac{2}{3^{n+1}}$. Suppose that they are at most ℓ levels apart in the stack, with I above J.*

(1) *For all $k \geq n$, we have*
$$\lambda(T^{h_k}(I) \cap I) \geq \frac{1}{3}\lambda(I).$$

(2) *There exists an integer H, depending on n and ℓ, such that*
$$\lambda(T^H(I) \cap J) \geq \left(\frac{1}{3}\right)^{\ell} \lambda(J).$$

Proof.

(1) For any level K in the nth iterate, let $K[0], K[1], K[2]$ be the three subintervals it's divided into in the $(n+1)^{\text{st}}$ iterate, from left to right. By the construction of T, we have
$$\begin{aligned} T^{h_n}(K[0]) &= K[1], \\ T^{h_n+1}(K[1]) &= K[2] \\ T^{h_n}(K[1]) &= (T^{-1}(K))[2] \end{aligned} \tag{8.2}$$
assuming that K isn't the bottom level. It follows that
$$\lambda(T^{h_n}(I) \cap I) \geq \frac{1}{3}\lambda(I), \qquad \lambda(T^{h_n}(I) \cap T^{-1}(I)) \geq \frac{1}{3}\lambda(I).$$
This proves part (1) for $k = n$.

If $k > n$, say with $k = n + j$, then I splits into 3^j subintervals in the kth iterate, and for each one of these pieces I', we have
$$\lambda(T^{h_k}(I') \cap I') \geq \frac{1}{3}\lambda(I').$$
Putting all the pieces together yields part (1).

(2) By (8.2), we have

$$\lambda(T^{h_n}(I) \cap T^{-1}(I)) \geq \frac{1}{3}\lambda(I) = \frac{1}{3}\lambda(T^{-1}(I)).$$

Since $T^{h_n}(I)$ contains two full levels in the $(n+1)^{\text{st}}$ iterate, applying (8.2) to each of them but with n replaced with $n+1$ implies that

$$\lambda(T^{h_n+h_{n+1}}(I) \cap T^{-r}(I)) \geq \left(\frac{1}{3}\right)^2 \lambda(T^{-r}(I))$$

for each $r \in \{0, 1, 2\}$. Similarly, since $T^{h_n+h_{n+1}}$ contains four full levels in the $(n+2)$nd iterate, if we apply (8.2) to all of them with n replaced with $n+2$, we have

$$\lambda(T^{h_n+h_{n+1}+h_{n+2}}(I) \cap T^{-r}(I)) \geq \left(\frac{1}{3}\right)^3 \lambda(T^{-r}(I))$$

for $r \in \{0, 1, 2, 3\}$. Continuing on in the same way, we find that $H = \sum_{i=0}^{\ell-1} h_{n+i}$ satisfies the hypothesis. ∎

Lemma 8.12 allows us to prove that the Chacón transformation is weakly mixing but not mixing.

Theorem 8.13. *The Chacón transformation is not mixing.*

Proof. For $n \geq 1$, if I is a level in the nth iterate of the Chacón transformation, then $\lambda(I) < \frac{1}{3}$. Then, for all $k \geq n$, we have

$$\lambda(T^{-h_k}(I) \cap I) = \lambda(T^{h_k}(I) \cap I) \geq \frac{1}{3}\lambda(I) > \lambda(I)^2.$$

Thus if $\lim_{n\to\infty} \lambda(T^{-n}(I) \cap I)$ exists, we must have

$$\lim_{n\to\infty} \lambda(T^{-n}(I) \cap I) \geq \frac{1}{3}\lambda(I) > \lambda(I)^2,$$

so it cannot be the case that $\lim_{n\to\infty} \lambda(T^{-n}(I) \cap I) = \lambda(I)^2$. Thus T is not mixing. ∎

Theorem 8.14. *The Chacón transformation is weakly mixing.*

Proof. It suffices to show that the Chacón transformation is doubly ergodic, so that's what we will do. Let A_1 and B_1 be two sets of positive measure, and find two intervals I_1 and J_1 in the same iterate of the construction such that

$$\lambda(A_1 \cap I_1) > \frac{2}{3}\lambda(I_1), \qquad \lambda(B_1 \cap J_1) > \frac{2}{3}\lambda(J_1).$$

Furthermore, we may do in such a way that I_1 is above J_1, say by ℓ levels. Let's suppose they are in the nth iterate of the construction. Let $\delta = \left(\frac{1}{3}\right)^{\ell}$. Now pick subintervals I and J of I_1 and J_1, respectively, such that

$$\lambda(A_1 \cap I) > \left(1 - \frac{\delta}{3}\right)\lambda(I), \qquad \lambda(B_1 \cap J) > \left(1 - \frac{\delta}{3}\right)\lambda(J).$$

Let $A = A_1 \cap I$ and $B = B_1 \cap I$.

Now, let $H = \sum_{i=0}^{\ell-1} h_{n+i}$. Then by Lemma 8.12 (2), we have

$$\lambda(T^H(I) \cap J) \geq \left(\frac{1}{3}\right)^{\ell} \lambda(J), \qquad \lambda(T^H(I) \cap I) \geq \left(\frac{1}{3}\right)^{\ell} \lambda(I).$$

Thus we have

$$\begin{aligned}
\lambda(T^H(A_1) \cap B_1) &\geq \lambda(T^H(A) \cap B) \\
&\geq \lambda(T^H(I) \cap J) - \lambda(I \setminus A) - \lambda(J \setminus B) \\
&\geq \delta\lambda(J) - \frac{\delta}{3}\lambda(I) - \frac{\delta}{3}\lambda(J) \\
&> 0
\end{aligned}$$

and

$$\begin{aligned}
\lambda(T^H(A_1) \cap A_1) &\geq \lambda(T^H(A) \cap A) \\
&\geq \lambda(T^H(I) \cap I) - 2\lambda(I \setminus A) \\
&> 0,
\end{aligned}$$

as desired. ■

8.5 Problems

(1) Show that the powers of 2 have density 0.

(2) Show that a union of finitely many density 0 sets is also a density 0 set.

(3) Let $I = \left(\frac{1}{3}, \frac{4}{9}\right)$, and let T be the Chacón transformation. Give an explicit description of a density 0 set D such that

$$\lim_{\substack{n\to\infty \\ n\notin D}} \lambda(T^{-n}(I) \cap I) = \lambda(I)^2.$$

Prove that your answer is correct.

(4) Let I and T be as in Problem 3. Let $J = \left(\frac{7}{9}, \frac{8}{9}\right)$. Give an explicit description of a density 0 set E such that

$$\lim_{\substack{n\to\infty \\ n\notin E}} \lambda(T^{-n}(I) \cap J) = \lambda(I)\lambda(J).$$

Prove that your answer is correct.

(5) Let T be the Bernoulli shift on $\{0,1\}^{\mathbb{N}}$, i.e. $T(x_1, x_2, \ldots) = (x_2, x_3, \ldots)$. Is T mixing? Weakly mixing? Prove that your answer is correct.

(6) Prove that a transformation T on a probability space $(X, \mathcal{A}, \mu)$ is mixing if and only if

$$\lim_{n\to\infty} \mu(T^{-n}(A) \cap A) = \mu(A)^2$$

for all $A \in \mathcal{A}$.

(7) Prove that the set D constructed in the proof of Proposition 8.6 has density 0.

(8) Show that Proposition 8.6 fails if the a_i's are not assumed to be all nonnegative.

(9) Let $a_1, a_2, \ldots$ be a bounded sequence of real numbers.

(a) Prove that

$$\lim_{n\to\infty} \frac{1}{n} \sum_{j=1}^{n} |a_j| = 0$$

if and only if there is a subset $D \subset \mathbb{N}$ of density 0 such that

$$\lim_{\substack{n\to\infty \\ n\notin D}} a_n = 0.$$

(b) Prove that

$$\lim_{n\to\infty} \frac{1}{n} \sum_{j=1}^{n} |a_j| = 0 \qquad \Longleftrightarrow \qquad \lim_{n\to\infty} \frac{1}{n} \sum_{j=1}^{n} a_j^2 = 0.$$

(10) Prove that if T is weakly mixing, then so is T^n for every positive integer n.

(11) (a) Does there exist a T that is not mixing, but where T^n is mixing for some n?
 (b) Does there exist a T that is not weakly mixing, but where T^n is weakly mixing for some n?
 (c) Does there exist a T that is weakly mixing but not mixing, but where T^n is mixing for some n?

(12) Prove that the dyadic odometer is not weakly mixing.

(13) Prove that a measure-preserving system $(X, \mathcal{A}, \mu, T)$ is doubly ergodic if and only if for all pairs of measurable sets A and B with positive measure, there is an integer $n > 0$ such that $T^{-n}(A) \cap A \neq \varnothing$ and $T^{-n}(A) \cap B \neq \varnothing$. (The definition of doubly ergodic is that there exists an n such that $\mu(T^{-n}(A) \cap A) > 0$ and $\mu(T^{-n}(A) \cap B) > 0$.)

(14) Let $(X, \mathcal{A}, \mu, T)$ be a measure-preserving system. Let $T \times T : X \times X \to X \times X$ be defined on $(X \times X, \mathcal{A} \times \mathcal{A}, \mu \times \mu)$ by $T \times T(x, y) = (T(x), T(y))$.
 (a) Prove that $T \times T$ is measure-preserving.
 (b) Prove that $T \times T$ is ergodic if and only if T is weakly mixing.
 (c) In terms of T, what does it mean for $T \times T$ to be weakly mixing? Mixing?
 (d) Find a transformation T such that $T \times T$ is not weakly mixing, or prove that no such T exists.

(15) A transformation T on a probability space is said to be *lightly mixing* if, for all measurable sets A and B of positive measure, there exists a number $r > 0$ and a positive integer N such that $\mu(T^{-n}(A) \cap B) \geq r$ for all $n \geq N$.
 (a) Prove that if T is ergodic, then T is lightly mixing if and only if for all measurable sets A of positive measure, there is an integer N such that $\mu(T^{-n}(A) \cap A) > 0$ for all integers $n \geq N$.
 (b) Prove that if T is lightly mixing, then it is weakly mixing.

(16) Determine whether the Chacón transformation is lightly mixing.

Chapter 9

Multiple Recurrence and Szemerédi's Theorem

9.1 A Brief Look at Additive Combinatorics

The subject of additive combinatorics is concerned with questions about additive structure in subsets of the integers. One of the main focal points of the subject is the study of arithmetic progressions. An *arithmetic progression* is a sequence of integers $a_1, a_2, \ldots, a_k$ such that there exists a positive integer d such that $a_i - a_{i-1} = d$ for $2 \leq i \leq k$. Equivalently, $a_1, a_2, \ldots, a_k$ is an arithmetic progression if there exist positive integers a and d such that $a_i = a + (i-1)d$ for $1 \leq i \leq k$. More precisely, we'll call such a sequence a *k-term arithmetic progression.*

Perhaps the first great result in additive combinatorics was van der Waerden's Theorem.

Theorem 9.1 (van der Waerden). *Given any two (not necessarily distinct) positive integers k and m and any coloring of the positive integers with m colors, then there is a monochromatic k-term arithmetic progression.*

For example, suppose $m = 2$ and $k = 3$. If we color all the integers one of two colors, say black or white, then there's a 3-term arithmetic progression that is all black or all white. In fact, we don't need to color all the positive integers; just the ones up to 9 will do. For instance, in the coloring

1	2	3	4	5	6	7	8	9

,

where the color is the foreground color, we have one monochromatic 3-term arithmetic progression, namely the white progression $1, 4, 7$.

More generally, we never need to color all the integers:

Theorem 9.2 (van der Waerden). *Let k and m be any two positive integers. Then there exists an integer $N = N(k, m)$ depending on k and m such that if we color the integers from 1 to N using m colors, then there is a monochromatic k-term arithmetic progression.*

This theorem is not that straightforward to prove, although it can be done in a purely elementary manner, with a bit of cleverness. See the rather remarkable book (actually a letter) [**Khi98**] for a beautiful proof of van der Waerden's Theorem.

The smallest such N is known as the *van der Waerden number*. We don't know very many exact values of N, but we do at least know that $N(3, 2) = 9$. The best we know in general is that

$$N(k, m) < 2^{2^{m^{2^{2^{k+9}}}}},$$

as shown in [**Gow01**]. This is believed to be *extremely* far from optimal, but this sort of thing is typical of our state of knowledge of additive combinatorics: we can only prove absurd bounds when the true answer should be far smaller.

Szemerédi's Theorem is a sort of *density* version of van der Waerden's Theorem. The idea is that if we're looking for some sort of pattern in one of several subsets of the integers, or the integers up to N, then maybe we're best off looking in the largest subset. In the case of van der Waerden's Theorem, we might ask whether there's some integer $N = N(k, m)$ such that if we color the integers from 1 to N using m colors, there's a monochromatic k-term arithmetic progression whose color is the one that's used most frequently. That seems intuitively plausible: we expect that it's easier to find a monochromatic k-term arithmetic progression in a color that's used often than in a color that's used only rarely. But van der Waerden's Theorem doesn't guarantee that we can do this, only that we can find a k-term arithmetic progression in *some* color, and we may be forced to look into the rarely used colors.

Well, if we're only going to be looking at the most common color, then we don't have to concern ourselves with coloring all the integers (or integers up to N); we only have to distinguish the subset of (say) red numbers. And since we're allowed to use any number of colors, we might as well assume that there are a lot of colors, so even the red numbers aren't that common; the only property that they need to have is that they should have *positive density*.

Definition 9.3. A subset S of the positive integers is said to have *positive upper density* if there is some $\varepsilon > 0$ such that

$$\frac{\#(S \cap \{1, 2, \ldots, n\})}{n} > \varepsilon$$

for infinitely many values of n.

In fact, even that is stronger than we really need: why should it need to be the *first* n integers, when any n consecutive numbers would do just as well? So, it would suffice if for every n, there is some m such that

$$\frac{\#(S \cap \{m, m+1, \ldots, m+n-1\})}{n} > \varepsilon, \tag{9.1}$$

a notion which is known as *positive upper Banach density*.

Here's some useful notation expressing the ideas we've just described:

Definition 9.4. Let S be a subset of the positive integers. The *upper density* $\bar{d}(S)$ is defined to be

$$\bar{d}(S) = \limsup_{n\to\infty} \frac{\#(S \cap \{1, 2, \ldots, n\})}{n}.$$

The *lower density* $\underline{d}(S)$ is defined to be

$$\underline{d}(S) = \liminf_{n\to\infty} \frac{\#(S \cap \{1, 2, \ldots, n\})}{n}.$$

The *upper Banach density* $d^*(S)$ is defined to be

$$d^*(S) = \limsup_{n\to\infty} \max_{m} \frac{\#(S \cap \{m, m+1, \ldots, m+n-1\})}{n},$$

and the *lower Banach density* $d_*(S)$ is defined to be

$$d_*(S) = \liminf_{n\to\infty} \min_{m} \frac{\#(S \cap \{m, m+1, \ldots, m+n-1\})}{n}.$$

Now we can state Szemerédi's Theorem.

Theorem 9.5 (Szemerédi). *If S is any subset of the positive integers with positive upper Banach density, then S contains arbitrarily long arithmetic progressions.*

By "containing arbitrarily long arithmetic progressions," we mean that, for every positive integer k, there is a k-term arithmetic progression in S. However, given (say) a k-term arithmetic progression, it isn't necessarily part of a $(k+1)$-term arithmetic progression; the $(k+1)$-term arithmetic progression might be somewhere else. Thus it need not be the case that S contains *infinitely* long arithmetic progressions.

A very important subset of the integers that does *not* have positive upper Banach density is the set of primes: it has density 0. However, there are still quite a lot of them, and we might wonder whether the primes contain arbitrarily long arithmetic progressions. And this is true, even though it doesn't follow from Szemerédi's Theorem. This is a fairly recent result, known as the *Green–Tao Theorem* [**GT08**].

There are several proofs of Szemerédi's Theorem. Szemerédi's original proof is completely elementary in the sense that it doesn't use any fancy results from elsewhere; however, it is complicated and intricate. Indeed, this is generally considered a hard result. The special case of 3-term arithmetic progressions was proven previously by Roth [**Rot53**].

We will see how to connect Szemerédi's Theorem to ergodic theory. This may seem rather surprising, since it doesn't seem to have anything to do with iterations. But it will be reminiscent of our proof of the Strong Law of Large Numbers using the Birkhoff Ergodic Theorem, putting the Bernoulli shift transformation on spaces of sequences. In order to prove it, we need a more powerful version of the Poincaré Recurrence Theorem.

9.2 The Furstenberg Multiple Recurrence Theorem

Recall that the Poincaré Recurrence Theorem says that if we have a measure-preserving system $(X, \mathcal{A}, \mu, T)$, then for any $A \in \mathcal{A}$ with $\mu(A) > 0$, there is some positive integer n such that $\mu(T^{-n}(A) \cap A) > 0$. In other words, A and *one* of its iterates intersect in a set of positive measure. But what can we say about A and *two* (or more) of its iterates? Well, of course we can say that there exist distinct m and n such that $\mu(T^{-m}(A) \cap T^{-n}(A) \cap A) > 0$, but then m and n aren't required to have anything to do with each other. But what if we require m to be $2n$? That is, A, its nth iterate, and its $(2n)^{\text{th}}$ iterate need to intersect in a set of positive measure.

This is the content of the Furstenberg Multiple Recurrence Theorem.

Theorem 9.6 (Furstenberg, [**Fur77**]). *Let $(X, \mathcal{A}, \mu, T)$ be a measure-preserving system. Then for any $A \in \mathcal{A}$ with $\mu(A) > 0$ and every positive integer k, there is a positive integer n such that*

$$\mu(A \cap T^{-n}(A) \cap T^{-2n}(A) \cap \cdots \cap T^{-(k-1)n}(A)) > 0.$$

Let's see how to prove Szemerédi's Theorem from the Furstenberg Multiple Recurrence Theorem.

Proof of Theorem 9.5. Let X be the space of binary sequences $(a_1, a_2, a_3, \dots)$, where each $a_i \in \{0, 1\}$. Let T be the left shift map

$$T(a_1, a_2, \dots) = (a_2, a_3, \dots).$$

We put a useful measure on X based on S from Szemerédi's Theorem, as follows. First, pick some $\varepsilon > 0$. Consider the sequence $s = (a_1, a_2, \dots) \in X$, where $a_i = 1$ if $i \in S$ and $a_i = 0$ if $i \notin S$. Now, for each positive integer n, find a positive integer m such that (9.1) holds, and then define the measure μ_n on X by setting

$$\mu_n = \frac{1}{n} \sum_{i=0}^{n-1} \delta_{T^{m+i}(s)}.$$

More concretely, for a measurable subset $A \subseteq X$, we look at the sequences $T^m(s), T^{m+1}(s), \dots, T^{m+n-1}(s)$ and see what proportion of these sequences are in A; that's what $\mu_n(A)$ is.

Now, recall our construction of invariant measures when discussing the Krylov–Bogolyubov Theorem: given any sequence of probability measures on a space, we can find a limit point in the weak* topology (see Appendix A), i.e. a probability measure μ on X and a subsequence

$k_1, k_2, \ldots$ of $\mathbb{N}$ such that $\lim_{n\to\infty} \mu_{k_n}(A) = \mu(A)$ for all measurable subsets A. (There may be many choices of μ.) Take one such limit measure μ.

This measure μ has the following properties:

- μ is T-invariant. To see this, let $A \in \mathcal{A}$. Then we have

$$\begin{aligned}\mu_{k_n}(T^{-1}(A)) - \mu_{k_n}(A) &= \frac{1}{k_n}\left(\sum_{i=0}^{k_n-1} \delta_{T^{m+i}(s)}(T^{-1}(A)) \right.\\ &\qquad \left. - \sum_{i=0}^{k_n-1} \delta_{T^{m+i}(s)}(A)\right)\\ &= \frac{1}{k_n}\left(\sum_{i=0}^{k_n-1} \delta_{T^{m+i+1}(s)}(A) \right.\\ &\qquad \left. - \sum_{i=0}^{k_n-1} \delta_{T^{m+i}(s)}(A)\right)\\ &= \frac{1}{k_n}(\delta_{T^{m+k_n}(s)}(A) - \delta_{T^m(s)}(A)),\end{aligned}$$

which tends to 0 as $n \to \infty$. Thus we have

$$\lim_{n\to\infty} (\mu_{k_n}(T^{-1}(A)) - \mu_{k_n}(A)) = 0,$$

so the limiting measure is T-invariant.

- Let $B = \{(a_1, a_2, \ldots) : a_1 = 1\}$. Then $\mu(B) > 0$, because $\mu_{k_n}(B) > \varepsilon$ for all n. Since $\mu(B) = \lim_{n\to\infty} \mu_{k_n}(B)$, it follows, that $\mu(B) \geq \varepsilon > 0$.

Now, we apply the Furstenberg Multiple Recurrence Theorem to the shift map T on X, with respect to the measure μ. Since $\mu(B) > 0$, for each positive integer k, there is a positive integer n such that

$$\mu(B \cap T^{-n}(B) \cap T^{-2n}(B) \cap \cdots \cap T^{-(k-1)n}(B)) > 0.$$

It follows that

$$\mu_{k_n}(B \cap T^{-n}(B) \cap \cdots \cap T^{-(k-1)n}(B)) > 0$$

for all sufficiently large n. Now, μ_{k_n} is an average of $\delta_{T^{m+i}(x)}$'s. So if $\mu_{k_n}(C) > 0$ for some measurable set C, it follows that some $\delta_{T^{m+i}(x)}(C) > 0$, i.e. that $T^{m+i}(x) \in C$ for some $m + i$ involved in the average.

In order for $\delta_{T^{m+i}(x)}(B \cap T^{-n}(B) \cap \cdots \cap T^{-(k-1)n}(B))$ to be positive, then, it must be the case that $T^{m+i}(x) \in B$, and $T^{m+i}(x) \in T^{-n}(B)$ (i.e. $T^{m+n+i}(x) \in B$), and so forth. The first condition, $T^{m+i}(x) \in B$ means that $a_{m+i+1} = 1$, i.e. $m + i + 1 \in S$. Similarly, the second condition, $T^{m+n+i}(x) \in B$ means that $m+n+i+1 \in S$, and so forth. Thus we find that S contains the k-term arithmetic progression $m + i + 1, m + i + 1 + n, m + i + 1 + 2n, \ldots, m + i + 1 + (k - 1)n$, and we're done! ∎

So, that's how the Furstenberg Multiple Recurrence Theorem implies Szemerédi's Theorem. We can also go the other way too and prove Furstenberg from Szemerédi. Let's see how that works. In fact, we'll be able to prove a slightly stronger version of Furstenberg's Theorem from Szemerédi:

Theorem 9.7 (Furstenberg). *Let $(X, \mathcal{A}, \mu, T)$ be a measure-preserving system. Then for all bounded nonnegative functions f on X with $\int_X f\,d\mu > 0$ and for all positive integers k, there is some nonnegative integer n such that*

$$\int_X \prod_{j=0}^{k-1} (f \circ T^{jn})\,d\mu > 0.$$

Note that we recover Theorem 9.6 by taking $f = \mathbb{1}_A$.

Proof of Theorem 9.7, assuming Szemerédi's Theorem. Let

$$B = \left\{ x \in X : f(x) > \frac{1}{2} \int_X f\,d\mu \right\}.$$

Then $\mu(B) > 0$, since if $\mu(B) = 0$, then we would have

$$\int_X f\,d\mu = \int_B f\,d\mu + \int_{B^c} f\,d\mu \leq 0 + \mu(B^c) \cdot \frac{1}{2} \int_X f\,d\mu < \int_X f\,d\mu.$$

For each positive integer n, define a function f_n to be

$$f_n(x) = \frac{1}{n} \sum_{j=0}^{n-1} \mathbb{1}_B(T^j(x)) = \frac{1}{n} \sum_{j=0}^{n-1} \mathbb{1}_{T^{-j}(B)}(x).$$

We have

$$\begin{aligned}
\int_X f_n\,d\mu &= \int_X \frac{1}{n}\sum_{j=0}^{n-1} \mathbb{1}_{T^{-j}(B)}\,d\mu \\
&= \frac{1}{n}\sum_{j=0}^{n-1}\int_X \mathbb{1}_{T^{-j}(B)}\,d\mu \\
&= \frac{1}{n}\sum_{j=0}^{n-1}\mu(T^{-j}(B)) \\
&= \frac{1}{n}\sum_{j=0}^{n-1}\mu(B) \\
&= \mu(B).
\end{aligned}$$

Note that $f_n \le 1$, so we may apply the reverse Fatou's Lemma, which gives

$$0 < \mu(B) = \limsup_{n\to\infty}\int_X f_n\,d\mu \le \int_X \limsup_{n\to\infty} f_n\,d\mu.$$

Let $g = \limsup_{n\to\infty} f_n$, and let $C = \{x \in X : g(x) > 0\}$. Thus $\mu(C) > 0$, because $\int_X g\,d\mu > 0$. For each $x \in C$, let $A(x) = \{j : T^j(x) \in B\}$. Since $x \in C$, we have

$$\begin{aligned}
0 &< g(x) \\
&= \limsup_{n\to\infty} f_n(x) \\
&= \limsup_{n\to\infty}\frac{1}{n}\sum_{j=0}^{n-1}\mathbb{1}_B(T^j(x)) \\
&= \bar{d}(A(x)).
\end{aligned}$$

Since $A(x)$ has positive upper density, Szemerédi's Theorem implies that it contains a k-term arithmetic progression, for each $x \in C$. We now partition C up into which arithmetic progression it contains: for positive integers a and s, let

$$C_{a,s} = \{x \in C : a, a+s, a+2s, \ldots, a+(k-1)s \in A(x)\}.$$

Since $C = \bigcup_{a,s=1}^{\infty} C_{a,s}$ is a countable partition of C and $\mu(C) > 0$, some $C_{a,s}$ must have positive measure as well. Then we have

$$\begin{aligned} 0 &< \mu(C_{a,s}) \\ &= \mu(C \cap T^{-a}(B) \cap T^{-a-s}(B) \cap \cdots \cap T^{-a-(k-1)s}(B)) \\ &= \mu(T^{-a}(B) \cap T^{-a-s}(B) \cap \cdots \cap T^{-a-(k-1)s}(B)) \\ &= \mu(T^{-a}(B \cap T^{-s}(B) \cap \cdots \cap T^{-(k-1)s}(B))) \\ &= \mu(B \cap T^{-s}(B) \cap \cdots \cap T^{-(k-1)s}(B)) \\ &= \int_X \prod_{j=0}^{k-1} \mathbb{1}_{T^{-js}(B)} \, d\mu. \end{aligned}$$

Finally, by the definition of B, we have

$$\mathbb{1}_B \le \frac{2f}{\int_X f \, d\mu},$$

which gives

$$0 < \int_X \prod_{j=0}^{k-1} \mathbb{1}_{T^{-js}(B)} \, d\mu \le \left(\frac{2}{\int_X f \, d\mu} \right)^k \int_X \prod_{j=0}^{k-1} (f \circ T^{js}) \, d\mu,$$

as desired. ■

There is another, even stronger, version of the Furstenberg Multiple Recurrence Theorem:

Theorem 9.8. *Let $(X, \mathcal{A}, \mu, T_1), \ldots (X, \mathcal{A}, \mu, T_k)$ be* commuting *measure-preserving systems, i.e. $T_i \circ T_j(x) = T_j \circ T_i(x)$ for all i, j with $1 \le i, j \le k$ and for all $x \in X$. Then for every $A \in \mathcal{A}$ with $\mu(A) > 0$, there is a positive integer n such that*

$$\mu(A \cap T_1^{-n}(A) \cap T_2^{-n}(A) \cap \cdots \cap T_k^{-n}(A)) > 0.$$

This is stronger, because we obtain the other form by letting $T_i = T^i$. These iterates of T all clearly commute with one another, but this version can also be used to prove a version of Szemerédi's Theorem for $\mathbb{Z}^d$, and this ergodic theory argument is the only proof known in this case. (One possible statement is that if $S \subseteq \mathbb{Z}^d$ is a subset with positive upper

Banach density (i.e. there exists some $\delta > 0$ such that there exist arbitrarily large d-dimensional cubes in $\mathbb{Z}^d$ for which at least δ proportion of the lattice points in the cube are in S), and $v_1, \ldots, v_k$ are any integer vectors in $\mathbb{Z}^d$, then there exist infinitely many $a \in \mathbb{Z}^d$ and $n \in \mathbb{N}$ such that $a, a + nv_1, a + nv_2, \ldots, a + nv_k \in S$.) If we drop the hypothesis that the T_i's commute, then the conclusion of Theorem 9.8 fails, as you will show in Problem 3.

Unfortunately for us, proving Furstenberg's Multiple Recurrence Theorem is beyond the scope of this book. So we will have to satisfy ourselves with our understanding of why Szemerédi's Theorem has something to do with ergodic theory, even though it doesn't appear to at first. In Chapter 10, we will prove an easier result of a similar flavor to Furstenbrerg's Multiple Recurrence Theorem, so you can see some of the ideas that go into the proof. It is worth pointing out that there is also an ergodic version of van der Waerden's Theorem, called the *Birkhoff Multiple Recurrence Theorem.* See [**Mul97**] for an easily readable account of the equivalence and a proof of the Birkhoff Multiple Recurrence Theorem.

9.3 The Furstenberg Correspondence Principle

As we saw, the question about arithmetic progressions in subsets of the integers is equivalent to a question about recurrences in measure-preserving transformations. This is just a special case of a much more general phenomenon. Let's illustrate the connection with another example. First, the ergodic theory version:

Theorem 9.9. *Let $(X, \mathcal{A}, \mu, T)$ be a measure-preserving system. Then, for any $A \in \mathcal{A}$ with $\mu(A) > 0$, there is some positive integer n such that $\mu(A \cap T^{-n^2}(A)) > 0$.*

Note the connection with Poincaré recurrence: this is saying that not just does a point of A eventually get back to A, but it does so in a perfect square number of steps. That might not be the *first* time it gets back, but it does get back in some square number of steps.

Now the number theory version:

Theorem 9.10. *Let S be a set of positive integers having positive upper Banach density. Then there exist distinct $s, t \in S$ such that $s - t$ is a perfect square.*

These are equivalent in much the same way as Szemerédi's Theorem and Furstenberg's Multiple Recurrence Theorem are equivalent. You will prove the equivalence in this case in Problem 14. Of course, either one of these results is hard to prove from scratch. We will see how to do it, and in fact how to prove a more general result known as the Polynomial Recurrence Theorem, in Chapter 10.

We now turn to the general Furstenberg Correspondence Principle, as presented in [**Fur81**].

Definition 9.11. Let $\mathcal{Q}$ be a collection of finite subsets of $\mathbb{N}$, and let $S \subseteq \mathbb{N}$ be an arbitrary subset. We say that S contains a *configuration* from $\mathcal{Q}$ if there exist a $Q \in \mathcal{Q}$ and an integer n such that $Q + n \subseteq S$.

In other words, S contains a translate of Q for some $Q \in \mathcal{Q}$. We can reinterpret Szemerédi's Theorem in this language by letting as follows: let $\mathcal{Q}_k$ be the set of arithmetic progressions of length k. Then Szemerédi's Theorem says that if $d^*(S) > 0$, then S contains a configuration from $\mathcal{Q}_k$ for every positive integer k. Similarly, Theorem 9.10 says that if $\mathcal{Q} = \{\{0, n^2\} : n \geq 1\}$, then if $d^*(S) > 0$, then S contains a configuration from $\mathcal{Q}$.

Theorem 9.12 (Furstenberg Correspondence Principle, [**Fur81**]). *Let $\mathcal{Q}$ be a family of configurations such that for any measure-preserving system $(X, \mathcal{A}, \mu, T)$ and any $A \in \mathcal{A}$ with $\mu(A) > 0$, there is a configuration $\{n_1, \ldots, n_k\} \in \mathcal{Q}$ such that*

$$\mu(T^{-n_1}(A) \cap T^{-n_2}(A) \cap \cdots \cap T^{-n_k}(A)) > 0.$$

Then if $S \subseteq \mathbb{N}$ with $d^*(S) > 0$, then S contains a configuration from $\mathcal{Q}$.

9.4 Arbitrarily Long Arithmetic Progressions

As we have discussed, any sequence with positive upper Banach density contains k-term arithmetic progressions for every k. But what about sets with Banach density 0? An important case of this is the sequence of prime numbers. The *Prime Number Theorem* says that $\pi(x) \approx \frac{x}{\log(x)}$, where $\pi(x)$ is the number of primes up to x, so that the upper and lower density of the primes is 0. (The upper Banach density of the primes is also 0, but this does not follow instantly from the Prime Number Theorem.) So we cannot use Szemerédi's Theorem to say anything about primes.

Fortunately, one of the most famous theorems of the 21st century (so far) answers this question.

Theorem 9.13 (Green–Tao, [**GT08**]). *The primes contain arbitrarily long arithmetic progressions.*

However, not just any infinite set will do. For instance, the squares do not contain arbitrarily long arithmetic progressions, as you will show in Problem 13. A conjecture of Erdős, mentioned in [**Erd81**], is the following:

Conjecture 9.14 (Erdős). *Let $A \subseteq \mathbb{N}$ be a set with $\sum_{a\in A} \frac{1}{a} = \infty$. Then A contains arbitrarily long arithmetic progressions.*

The primes are an example of such a sequence, since it is a famous result of Euler that $\sum_{p \text{prime}} \frac{1}{p} = \infty$. But in general, Erdős's Conjecture is still open; Green and Tao's proof relies on special properties of the primes not shared by other sets A with $\sum_{a\in A} \frac{1}{a} = \infty$.

The converse of Erdős's Conjecture is false; see Problem 5.

9.5 Schur's Work on Arithmetic Progressions

How did van der Waerden's Theorem come about? What caused people to wonder about such things? One of the people before van der Waerden who was interested in such questions was Schur. He proved the following theorem:

Theorem 9.15 (Schur, [**Sch16**]). *Let m be a positive integer, and let $N > m!\,e$. If the integers $1, 2, \ldots, N$ are colored using m colors, then there are integers a, b, c with the same color such that $a + b = c$.*

To prove this, we first need a special case of Ramsey theory:

Theorem 9.16 (Greenwood and Gleason, [**GG55**]). *For any integer m, there exists an integer $N = N(m)$ such that any m-coloring of the complete graph K_N contains a monochromatic triangle.*

Proof. We proceed by induction on m. The base case $m = 1$ is trivial (take $N = 3$). Suppose now that the result is true for m, and we'll prove it for $m+1$. Suppose that any m-coloring of K_N contains a monochromatic triangle. Now, let $N(m+1) = (m+1)(N(m)-1)+2$. Assume that the edges of $K_{N(m+1)}$ are $(m+1)$-colored, and let v_0 be a vertex of $K_{N(m+1)}$. Among the $N(m+1)-1 = (m+1)(N(m)-1)+1$ neighbors of v_0, there is some color, say red, such that at least $N(m)$ edges containing v_0 are red, by the pigeonhole principle. If any two of them are connected by a red edge, we have a red triangle, and we're done. If not, then they form a $K_{N(m)}$ colored in m colors: all the colors other than red. By the inductive hypothesis, they contain a monochromatic triangle. ∎

In fact, we can give a nice bound for $N(m)$ easily enough: by the recurrence, we have $N(m) \leq \lfloor m!\, e \rfloor + 1$.

Proof of Schur's Theorem. Construct the graph K_N on vertex set $\{1, 2, \ldots, N\}$, and color the edge from i to j using color k if $|i - j|$ is colored using color k. By Theorem 9.16, this graph contains a monochromatic triangle, say with vertices x, y, z where $x < y < z$. Let $a = y - x$, $b = z - y$, and $c = z - x$. This means that a, b, c are the same color, and $a + b = c$, so we're done. ∎

See Problem 11 for a slight strengthening of Schur's Theorem.

Schur followed up with a few related questions:

Question 9.17. *For a positive integer k and a sufficiently large prime p, do there exist k consecutive quadratic residues modulo p? Do there exist k consecutive nonresidues modulo p?*

Schur conjectured that the answers to both of these questions was "yes," and indeed they are. Nowadays these problems can be solved fairly easily with some standard techniques in analytic number theory. But Schur didn't have these available to him a century ago, and so he thought that the following less structured conjecture might be helpful:

Conjecture 9.18. *For any positive integer k, there is an integer N depending on k such that whenever the integers* $1, 2, \ldots, N$ *are partitioned into two disjoint subsets, one of these subsets contains a k-term arithmetic progression.*

This, of course, is the $m = 2$ case of van der Waerden's Theorem.

That's just a snippet of the history of this problem. For more of the story, as well as many other related (and unrelated) coloring problems, see the delightful book [**Soi09**].

9.6 Problems

(1) Prove that for any $S \subseteq \mathbb{N}$, we have $d_*(S) \leq \underline{d}(S) \leq \bar{d}(S) \leq d^*(S)$.

(2) Prove that Szemerédi's Theorem implies van der Waerden's Theorem.

(3) Give an example to show that the conclusion of Theorem 9.8 fails if $T_1, \ldots, T_k$ do not commute.

(4) Prove that Szemerédi's Theorem is equivalent to the following finitary version: For every $\varepsilon > 0$ and every positive integer k, there is an integer n depending on ε and k such that for any subset S of $\{1, 2, \ldots, n\}$ of size at least εn, there is a k-term arithmetic progression in S.

(5) Prove that the converse of Conjecture 9.14 is false, in that there is a set A with $\sum_{a \in A} \frac{1}{a} < \infty$, but A contains arbitrarily long arithmetic progressions.

(6) Let $\alpha \in \mathbb{R} \setminus \mathbb{Q}$, and let $A = \left(0, \frac{1}{100}\right) \subseteq \mathbb{R}/\mathbb{Z}$. For each positive integer k, prove directly (i.e. without using the Furstenberg Multiple Recurrence Theorem or Szemerédi's Theorem, or any other theorem that you do not know how to prove from scratch) that there exists an n (in terms of k and α) such that

$$\lambda(A \cap R_{\alpha}^{-n}(A) \cap R_{\alpha}^{-2n}(A) \cap \cdots \cap R_{\alpha}^{-(k-1)n}(A)) > 0.$$

Can you find a bound on how large n has to be?

(7) Let $A = \left(0, \frac{1}{100}\right) \subseteq \mathbb{R}/\mathbb{Z}$. Find an n such that

$$\lambda(A \cap R_{\sqrt{2}}^{-n}(A) \cap R_e^{-n}(A) \cap R_\pi^{-n}(A)) > 0.$$

(8) Prove directly that the Furstenberg Multiple Recurrence Theorem holds for weakly mixing transformations on a probability space.

(9) Using van der Waerden's Theorem, but *not* Szemerédi's Theorem, prove the following theorem of Kakeya and Morimoto: suppose that that $a_1 < a_2 < \cdots$ is a sequence of positive integers such that there exists an integer M such that $a_1 \leq M$ and $a_{n+1} - a_n \leq M$ for all $n \geq 1$. Then the sequence $a_1, a_2, \ldots$ contains arbitrarily long arithmetic progressions.

(10) With the hypotheses of Problem 9, show that $a_1, a_2, \ldots$ need not contain an *infinite* arithmetic progression by giving an example of such a sequence with no infinite arithmetic progressions.

(11) Prove the following strengthening of Schur's Theorem: for any positive integer m, there exists a positive integer n depending on m such that if $N > n$, then any m-coloring of $\{1, 2, \ldots, N\}$ contains a triple of *distinct* integers a, b, c where $a + b = c$. (The only changes are that now a, b, c are required to be distinct, and that the bound on N may be larger than $m!\,e$.)

(12) Prove *Varnavides's Theorem*, which says that if $\delta > 0$ and $k \geq 1$, then there exists an integer $N(k, \delta)$ such that if $n > N(k, \delta)$, then any subset $S \subseteq \{1, 2, \ldots, n\}$ of cardinality at least δn contains at least $\frac{n^2}{N(k,\delta)}$ arithmetic progressions of length k.

(13) Prove that the perfect squares do not contain arbitrarily long arithmetic progressions. What is the longest arithmetic progression of perfect squares?

(14) Prove that Theorems 9.9 and 9.10 are equivalent, in the same way that Szemerédi's Theorem and Furstenberg's Multiple Recurrence Theorem are equivalent.

(15) Say that a set $S \subseteq \mathbb{N}$ of positive integers is a *set of recurrence* if for any measure-preserving system $(X, \mathcal{A}, \mu, T)$, if $A \in \mathcal{A}$ has $\mu(A) > 0$,

then there is some $s \in S$ such that $\mu(A \cap T^{-s}(A)) > 0$. Prove that if S_1 and S_2 are not sets of recurrence, then neither is $S_1 \cup S_2$.

(16) Let $a_1 < a_2 < a_3 < \cdots$ be an infinite sequence of positive integers. We call the sequence *lacunary* if there is some $q > 1$ such that $\frac{a_{n+1}}{a_n} > q$ for all $n \geq 1$. Prove that a lacunary sequence is not a set of recurrence.

Chapter 10

Polynomial Recurrence

10.1 Polynomial Recurrence

The proof of Furstenberg's Multiple Recurrence Theorem is beyond the scope of this book, but see [**EW11**, Chapter 7] for a readable account of it. In this chapter, we will look at another result that is of a similar flavor to the Furstenberg Multiple Recurrence Theorem, whose proof incorporates some of the same ideas, but that is easier. This is the Polynomial Recurrence Theorem, due (independently) to Furstenberg [**Fur77**] and Sárközy [**Sár78**].

Theorem 10.1 (Furstenberg, Sárközy). *If S is any subset of the positive integers with positive upper Banach density, and $p(x)$ is any polynomial with integer coefficients and $p(0) = 0$, then there exist integers $x, y \in S$ and a positive integer n such that $x - y = p(n)$.*

Note that Theorem 9.10 is the special case of $p(x) = x^2$. As before, we can use the Furstenberg Correspondence Principle to convert this into a result in ergodic theory. The corresponding result is the following, called the *Polynomial Recurrence Theorem*:

Theorem 10.2. *If $p(x)$ is a polynomial with integer coefficients and $p(0) = 0$, then for every measure-preserving system $(X, \mathcal{A}, \mu, T)$ and every measurable set $A \in \mathcal{A}$ with $\mu(A) > 0$, there exists a positive integer n such that $\mu(A \cap T^{-p(n)}(A)) > 0$.*

We can use the Furstenberg Correspondence Principle (Theorem 9.12) to show that Theorem 10.2 implies Theorem 10.1. This follows from taking $\mathcal{Q} = \{\{0, p(n)\} : n \geq 1\}$ in Theorem 9.12.

We will be able to give a proof of Theorem 10.2, modulo a few technical details. Our proof will follow the treatment in [**EW11**, §7.4, pp. 180–184]; see that book to see how all the details can be filled in.

The basic idea behind the proof is that we wish to split a certain space of functions on X into two pieces: a chaotic part and an orderly part. Then we will show that it is sufficient to treat the two parts separately, then address each one.

10.2 Chaotic and Orderly Parts

Let $(X, \mathcal{A}, \mu, T)$ be a measure-preserving system. Let L^2 (or $L^2(\mu)$) denote the space of square-integrable functions on X, i.e. the set of measurable $f : X \to \mathbb{R}$ such that $\int_X f^2 \, d\mu < \infty$. (We can also consider complex-valued functions, with slight changes, but this will not be necessary for our main result, so we stick to the slightly simpler case.) For $f \in L^2$, let

$$\|f\|_2 = \left(\int_X f^2 \, d\mu \right)^{1/2}.$$

For $f \in L^2$, let $U_T f$ be the function given by $U_T f(x) = f(T(x))$. As usual, we will consider functions to be the same if they agree except on a null set.

For each positive integer a, let

$$\mathcal{H}_a = \{f \in L^2 : U_T^a f = f\},$$

i.e. $\mathcal{H}_a$ consists of those functions f for which $f(T^a(x)) = f(x)$ for almost all $x \in X$. Then let

$$\mathcal{V}_a = \left\{ f \in L^2 : \lim_{N\to\infty} \left\| \frac{1}{N} \sum_{n=0}^{N-1} U_T^{an} f \right\|_2 = 0 \right\}.$$

Any function $f \in L^2$ can be written uniquely as a sum of a function in $\mathcal{H}_a$ and a function in $\mathcal{V}_a$, as you will show in Problem 1. We write $L^2 = \mathcal{H}_a \oplus \mathcal{V}_a$ to express this. Furthermore, this sum is *orthogonal*, meaning that if $f \in \mathcal{H}_a$ and $g \in \mathcal{V}_a$, then

$$\int_X fg\,d\mu = 0.$$

Recall that T is ergodic if and only if the only T-invariant functions are constants, so $\mathcal{H}_a$ always contains the constant functions, and it contains other functions as well precisely when T^a is not ergodic. We think of $\mathcal{H}_a$ as being the orderly part, where T^a is periodic, and $\mathcal{V}_a$ is the chaotic part, where averages of iterates of T^a exhibit as much cancellation as possible.

Definition 10.3. We say that T is *totally ergodic* if T^a is ergodic for all positive integers a.

Definition 10.4. The *rational spectrum* of T is

$$\mathcal{H}_{\text{rat}} = \overline{\{f \in L^2 : U_T^a f = f \text{ for some } a \geq 1\}}.$$

The bar denotes the closure, i.e. all possible limits (in the L^2-norm) of functions that are invariant under T^a for some a. It is not too difficult to check that $\mathcal{H}_{\text{rat}}$ is a subspace of L^2.

Definition 10.5. The *totally ergodic component* of T is

$$\mathcal{V}_* = \left\{ f \in L^2 : \lim_{N\to\infty} \left\| \frac{1}{N} \sum_{n=0}^{N-1} U_T^{an} f \right\|_2 = 0 \text{ for all } a \geq 1 \right\}.$$

Here are some basic properties of the rational spectrum and the totally ergodic component:

Proposition 10.6.

- *$\bigcup_{a=1}^{\infty} \mathcal{H}_a$ is dense in $\mathcal{H}_{\text{rat}}$.*
- *$\bigcap_{a=1}^{\infty} \mathcal{V}_a = \mathcal{V}_*$.*
- *$\mathcal{H}_{\text{rat}}^{\perp} = \mathcal{V}_*$, i.e.*

$$\mathcal{V}_* = \left\{ f \in L^2 : \int_X fg\,d\mu = 0 \textit{ for all } g \in \mathcal{H}_{\text{rat}} \right\}.$$

- *$L^2 = \mathcal{H}_{\text{rat}} \oplus \mathcal{V}_*$.*
- *$\mathcal{H}_{\text{rat}}$ and $\mathcal{V}_*$ are T-invariant.*

You will prove Proposition 10.6 in Problem 2.

Since every function in L^2 can be written uniquely as the sum of a function in $\mathcal{H}_{\text{rat}}$ and a function in $\mathcal{V}_*$ (and of a function in $\mathcal{H}_a$ and a function in $\mathcal{V}_a$ for each a), this is true of $\mathbb{1}_A$ for any $A \in \mathcal{A}$. Let us write the decomposition of $\mathbb{1}_A$ as $\mathbb{1}_A = f + g$ where $f \in \mathcal{H}_{\text{rat}}$ and $g \in \mathcal{V}_*$, and similarly $\mathbb{1}_A = f_a + g_a$ where $f_a \in \mathcal{H}_a$ and $g_a \in \mathcal{V}_a$. We have

$$\int_X \mathbb{1}_A \, d\mu = \mu(A) = \int_X f \, d\mu + \int_X g \, d\mu = \int_X f_a \, d\mu + \int_X g_a \, d\mu.$$

Now, because T is measure-preserving, we have

$$\int_X U_T^{an} h \, d\mu = \int_X h \, d\mu,$$

so it follows that $\int_X g_a \, d\mu = 0$. Since g is the limit of functions in $\mathcal{V}_a$ for various values of a, it also follows that $\int_X g \, d\mu = 0$. Thus we have

$$\mu(A) = \int_X f \, d\mu = \int_X f_a \, d\mu.$$

Lemma 10.7. $f_a, f \geq 0$ *almost everywhere.*

Proof. We begin with f_a. Let $B = \{x \in X : f_a(x) < 0\}$. Because f_a is T^a-invariant, if $x \in B$, then $T^a(x) \in B$, so it follows that B itself is T^a-invariant. Thus $f_a \mathbb{1}_B \in \mathcal{H}_a$, and similarly, $g_a \mathbb{1}_B \in \mathcal{V}_a$. Thus the decomposition of $\mathbb{1}_{A \cap B}$ is $f_a \mathbb{1}_B + g_a \mathbb{1}_B$, where $f_a \mathbb{1}_B \in \mathcal{H}_a$ and $g_a \mathbb{1}_B \in \mathcal{V}_a$. Thus $\int_X g_a \mathbb{1}_B \, d\mu = 0$, so $\int_X f_a \mathbb{1}_B = \mu(A \cap B) \geq 0$. But $f_a \mathbb{1}_B$ is nonpositive, so it must zero almost everywhere, i.e. $\mu(B) = 0$. Since we are only considering functions up to null sets, this shows that $f_a \geq 0$.

Now, $f = \lim_{a \to \infty} f_{a!}$ (see Problem 3), so we have $f \geq 0$ as well. ■

We will show a stronger statement than polynomial recurrence: we will show that

$$\lim_{N \to \infty} \frac{1}{N} \sum_{n=0}^{N-1} \mu(A \cap T^{p(n)}(A))$$

exists and is positive. Note that we have

$$\begin{aligned}\mu(A \cap T^{p(n)}(A)) &= \int_X \mathbb{1}_A \mathbb{1}_{T^{p(n)}(A)}\, d\mu \\ &= \int_X (f+g)(U_T^{p(n)} f + U_T^{p(n)} g)\, d\mu \\ &= \int_X f \cdot U_T^{p(n)} f\, d\mu + \int_X g \cdot U_T^{p(n)} g\, d\mu,\end{aligned}$$

because the integral of a product of a function in $\mathcal{H}_{\text{rat}}$ and a function in $\mathcal{V}_*$ is 0. Thus it suffices to work with the $\mathcal{H}_{\text{rat}}$ part and the $\mathcal{V}_*$ part separately. We will show that the average of the first component is positive, while the average of the second component is zero, which implies the result.

10.3 $\mathcal{H}_{\text{rat}}$

We now tackle the rational spectrum. We show that

$$\lim_{N\to\infty} \frac{1}{N} \sum_{n=0}^{N-1} \int_X f \cdot U_T^{p(n)} f\, d\mu > 0.$$

Note that the values of $p(n)$ modulo a are periodic with period a (or some divisor of a), so if $h \in \mathcal{H}_a$, then the sequence of functions $U_T^{p(n)} h$ is periodic with period a. Thus for $h \in \mathcal{H}_a$, we have

$$\lim_{N\to\infty} \frac{1}{N} \sum_{n=0}^{N-1} \int_X h \cdot U_T^{p(n)} h\, d\mu = \frac{1}{a} \sum_{n=0}^{a-1} \int_X h \cdot U_T^{p(n)} h\, d\mu.$$

In particular, the limit exists. Since this limit exists for functions on a dense subset of $\mathcal{H}_{\text{rat}}$, namely $\bigcup_{a=1}^{\infty} \mathcal{H}_a$, it holds on all of $\mathcal{H}_{\text{rat}}$.

We must now show that, for f as above in the decomposition of $\mathbb{1}_A$, the limit is positive. Let $\varepsilon = \min(\frac{1}{6\|f\|_2}\mu(A)^2, \|f\|_2)$, and choose $a \geq 1$ such that $\|f - f_a\|_2 < \varepsilon$; note that $\|f_a\|_2 < 2\|f\|_2$. For all $n \geq 1$, $p(an)$ is divisible by a since $p(0) = 0$, so $U_T^{p(an)} f_a = f_a$. Thus we have

$$\int_X f_a \cdot U_T^{p(an)} f_a\, d\mu = \int_X f_a^2\, d\mu \geq \left(\int_X f_a\, d\mu\right)^2 = \mu(A)^2$$

by the Cauchy–Schwarz Inequality. Thus we have

$$\begin{aligned}
\int_X f \cdot U_T^{p(an)} f \, d\mu &= \int_X f_a^2 \, d\mu + \int_X f_a \cdot (U_T^{p(an)} f - f_a) \, d\mu \\
&\quad + \int_X (f - f_a) \cdot U_T^{p(an)} f \, d\mu \\
&= \int_X f_a^2 \, d\mu + \int_X f_a \cdot (U_T^{p(an)} f - U_T^{p(an)} f_a) \, d\mu \\
&\quad + \int_X (f - f_a) \cdot U_T^{p(an)} f \, d\mu \\
&\geq \mu(A)^2 - \|f_a\|_2 \cdot \|U_T^{p(an)} (f - f_a)\|_2 \\
&\quad - \|f - f_a\|_2 \cdot \|U_T^{p(an)} f\|_2 \\
&= \mu(A)^2 - \|f_a\|_2 \cdot \|f - f_a\|_2 - \|f - f_a\|_2 \cdot \|f\|_2 \\
&\geq \frac{1}{2} \mu(A)^2 \\
&> 0.
\end{aligned}$$

Since $f \geq 0$ almost everywhere, $U_T^{p(n)} f \geq 0$ almost everywhere as well, so it follows that

$$\begin{aligned}
\lim_{N\to\infty} \frac{1}{N} \sum_{n=0}^{N-1} \int_X f \cdot U_T^{p(n)} f \, d\mu &\geq \lim_{N\to\infty} \frac{1}{N} \sum_{n=0}^{N-1} \int_X f \cdot U_T^{p(an)} f \, d\mu \\
&\geq \frac{1}{2a} \mu(A)^2 \\
&> 0.
\end{aligned}$$

This completes the proof for $\mathcal{H}_{\text{rat}}$.

10.4 $\mathcal{V}_*$

We now have to deal with the totally ergodic component and show that the average is zero there, i.e.

$$\lim_{N\to\infty} \frac{1}{N} \sum_{n=0}^{N-1} \int_X g \cdot U_T^{p(n)} g \, d\mu = 0.$$

In fact, we will show that

$$\lim_{N\to\infty}\left\|\frac{1}{N}\sum_{n=0}^{N-1}U_T^{p(n)}g\right\|_2=0,$$

which implies the result thanks to the Cauchy–Schwarz Inequality:

$$\left|\int_X g\cdot\frac{1}{N}\sum_{n=0}^{N-1}U_T^{p(n)}g\,d\mu\right|\le\|g\|_2\cdot\left|\int_X\frac{1}{N}\sum_{n=0}^{N-1}U_T^{p(n)}g\,d\mu\right|.$$

Since $\|g\|_2$ is constant and the second term on the right tends to 0, this implies that

$$\lim_{N\to\infty}\frac{1}{N}\sum_{n=0}^{N-1}\int_X g\cdot U_T^{p(n)}g\,d\mu=0.$$

In order to do this, we need to use a key lemma that is frequently used for these sorts of things: the van der Corput Lemma. See [**vdC31**] for the original paper, and [**BM16**] for a survey on various versions and applications. Here is the form that we will need:

Lemma 10.8 (van der Corput). *Let $h_0, h_1, \ldots$ be a sequence of functions in $L^2(\mu)$, such that there exists some real number B with $\|h_n\|_2\le B$ for all n. If for every positive integer d, we have*

$$\lim_{N\to\infty}\sum_{n=0}^{N-1}\int_X h_{n+d}h_n\,d\mu=0,$$

then

$$\lim_{N\to\infty}\frac{1}{N}\sum_{n=0}^{N-1}h_n=0$$

almost everywhere.

Proof. Let $\varepsilon>0$. For any fixed positive integer D, we have

$$\left\|\frac{1}{N}\sum_{n=0}^{N-1}h_n-\frac{1}{D}\sum_{d=1}^{D}\frac{1}{N}\sum_{n=0}^{N-1}h_{n+d}\right\|_2<\frac{\varepsilon}{2}$$

for sufficiently large N, because the terms in the two sums are mostly the same and the $\|h_n\|_2$'s are bounded. Thus it suffices to show that for

D sufficiently large, we have

$$\limsup_{N\to\infty} \left\| \frac{1}{D} \sum_{d=1}^{D} \frac{1}{N} \sum_{n=0}^{N-1} h_{n+d} \right\|_2 < \frac{\varepsilon}{2}.$$

By the Cauchy–Schwarz Inequality and the convexity of the squaring function, we have

$$\begin{aligned}
\limsup_{N\to\infty} \left\| \frac{1}{N} \sum_{n=0}^{N-1} \frac{1}{D} \sum_{d=1}^{D} h_{n+d} \right\|_2^2 &\le \limsup_{N\to\infty} \left(\frac{1}{N} \sum_{n=0}^{N-1} \left\| \frac{1}{D} \sum_{d=1}^{D} h_{n+d} \right\|_2 \right)^2 \\
&\le \limsup_{N\to\infty} \frac{1}{N} \sum_{n=0}^{N-1} \left\| \frac{1}{D} \sum_{d=1}^{D} h_{n+d} \right\|_2^2 \\
&= \limsup_{N\to\infty} \frac{1}{N} \sum_{n=1}^{N} \frac{1}{D^2} \sum_{d,e=1}^{D} \int_X h_{n+d} h_{n+e} \, d\mu \\
&\le \frac{1}{D^2} \sum_{d,e=1}^{D} \limsup_{N\to\infty} \frac{1}{N} \sum_{n=0}^{N-1} \int_X h_{n+d} h_{n+e} \, d\mu \\
&\le \frac{1}{D^2} \sum_{d,e=1}^{D} \limsup_{N\to\infty} \left| \frac{1}{N} \sum_{n=0}^{N-1} \int_X h_{n+d} h_{n+e} \, d\mu \right|.
\end{aligned}$$

Now, for any pair (d, e) with $1 \le d, e \le D$ and $d \ne e$, we have

$$\limsup_{N\to\infty} \left| \frac{1}{N} \sum_{n=0}^{N-1} \int_X h_{n+d} h_{n+e} \, d\mu \right| = \lim_{N\to\infty} \sum_{n=0}^{N-1} \int_X h_{n+|d-e|} h_n \, d\mu = 0.$$

Thus we have

$$\limsup_{N\to\infty} \left\| \frac{1}{N} \sum_{n=0}^{N-1} \frac{1}{D} \sum_{d=1}^{D} h_{n+d} \right\|_2^2 = 0,$$

as desired. ■

We now return to showing that

$$\left\| \lim_{N\to\infty} \sum_{n=0}^{N-1} U_T^{p(n)} g \right\|_2 = 0.$$

We do this by induction on $\deg p$. We will not assume that $p(0) = 0$ for this part, as this will not persist through the inductive step, so our base case, with p a linear polynomial, must prove it for polynomials of the form $p(n) = an + b$. Since $g \in \mathcal{V}_*$, hence in $\mathcal{V}_a$, we have

$$\lim_{N\to\infty} \left\| \frac{1}{N} \sum_{n=0}^{N-1} U_T^{an} g \right\|_2 = 0,$$

and since $\mathcal{V}_*$ is T-invariant, $T^b g \in \mathcal{V}_*$ as well. Thus we have

$$\lim_{N\to\infty} \left\| \frac{1}{N} \sum_{n=0}^{N-1} U_T^{an+b} g \right\|_2 = 0$$

as well. This completes the proof for linear polynomials.

Now suppose that the result holds for all polynomials of degree less than k, and that $\deg p = k$. Let $h_n = U_T^{p(n)} g$ in van der Corput's Lemma. We have

$$\begin{aligned} \int_X h_{n+d} h_n \, d\mu &= \int_X U_T^{p(n+d)} g \cdot U_T^{p(n)} g \, d\mu \\ &= \int_X g \cdot U_T^{p(n+d)-p(n)} g \, d\mu. \end{aligned}$$

Now, if we let $q(n) = p(n+d) - p(n)$, then q is a polynomial of degree $k-1$. Let $v_n = U_T^{q(n)} g$. By the inductive hypothesis, we have

$$\begin{aligned} \lim_{N\to\infty} \frac{1}{N} \left\| \sum_{n=0}^{N-1} \int_X h_{n+d} h_n \, d\mu \right\|_2 &= \lim_{N\to\infty} \frac{1}{N} \left\| \sum_{n=1}^{N} \int_X g \cdot U_T^{q(n)} g \, d\mu \right\|_2 \\ &= 0. \end{aligned}$$

Thus by the van der Corput Lemma, we have

$$\lim_{N\to\infty} \frac{1}{N} \sum_{n=0}^{N-1} h_n = 0$$

in L^2, i.e.

$$\lim_{N\to\infty} \left\| \frac{1}{N} \sum_{n=0}^{N-1} U_T^{p(n)} g \right\|_2 = 0.$$

This completes the proof of the totally ergodic case, hence of the Polynomial Recurrence Theorem.

10.5 Problems

(1) Show that $L^2 = \mathcal{H}_a \oplus \mathcal{V}_a$.

(2) Prove Proposition 10.6.

(3) Show that if $h \in L^2$ decomposes and $f_a + g_a$ in $\mathcal{H}_a \oplus \mathcal{V}_a$ and $f + g$ in $\mathcal{H}_{\text{rat}} \oplus \mathcal{V}_*$, then $f = \lim_{a\to\infty} f_{a!}$.

(4) Show that the polynomial recurrence theorem fails if we drop the hypothesis that $p(0) = 0$. Can it be replaced by a weaker condition?

(5) Prove the following Cesàro version of the van der Corput Lemma: With notation as in the version in the chapter, suppose that

$$\lim_{D\to\infty} \frac{1}{D} \sum_{d=1}^{D} \limsup_{N\to\infty} \left| \frac{1}{N} \sum_{n=0}^{N-1} \int_X h_{n+d} h_n \, d\mu \right| = 0.$$

Then

$$\lim_{N\to\infty} \frac{1}{N} \sum_{n=0}^{N-1} h_n = 0.$$

(6) We say that a sequence $x_0, x_1, x_2, \ldots$ in $\mathbb{R}/\mathbb{Z}$ is *equidistributed* if whenever $0 \le a < b \le 1$, we have

$$\lim_{N\to\infty} \frac{\#\{n : 0 \le n < N, x_n \in (a,b)\}}{N} = b - a.$$

Show that if α is irrational and $x_n = \{n\alpha\}$, then x_n is equidistributed. Here $\{x\}$ denotes the fractional part of x.

(7) Show that $x_n = \{\sqrt{n}\}$ is equidistributed.

(8) Show that $x_n = \{\log n\}$ is not equidistributed.

(9) Prove *Weyl's criterion*, which says the following. Let x_n be a sequence in $\mathbb{R}/\mathbb{Z}$. Then the following are equivalent:

(a) x_n is equidistributed.

(b) The sequence of measures

$$\mu_N = \frac{1}{N} \sum_{n=0}^{N-1} \delta_{x_n}$$

converges to the Lebesgue measure λ in the weak* topology. (See Appendix A for a definition and properties.) This means that for all Lebesgue-measurable functions f on $\mathbb{R}/\mathbb{Z}$, we have

$$\lim_{N\to\infty} \int_{\mathbb{R}/\mathbb{Z}} f\, d\mu_N = \int_{\mathbb{R}/\mathbb{Z}} f\, d\lambda.$$

(c) For every positive integer k, we have

$$\lim_{N\to\infty} \frac{1}{N} \sum_{n=0}^{N-1} e^{2\pi i k x_n} = 0.$$

(10) Suppose that x_n is a sequence in $\mathbb{R}/\mathbb{Z}$ such that for every m, the sequence $y_n = x_{n+m} - x_n$ is equidistributed. Show that x_n is equidistributed.

(11) Show that if $p(x)$ is a polynomial such that at least one nonconstant coefficient is irrational, then the sequence $x_n = \{p(n)\}$ is equidistributed.

Chapter 11

Entropy

11.1 Introduction to Entropy

The idea of entropy, first introduced by Shannon in [**Sha48**], is to measure the level of surprise one gets from additional information. For instance, suppose I'm thinking of a number $\alpha \in [0, 1)$. If I tell you that the first digit of α is a 4, i.e. $\alpha \in [.4, .5)$, then you have ten times as much information about α as you did just knowing that $\alpha \in [0, 1)$. If I further reveal that the second digit of α is 8, i.e. $\alpha \in [.48, .49)$, then you have 100 times as much information about α as you did to begin with.

We can also present information about α in other ways. For example, suppose I tell you what the first term in the continued fraction expansion of α is. How much information do you get then? This is more subtle, because it depends on what the term is. If the first continued fraction term is a 1, then this says that $\alpha \in \left(\frac{1}{2}, 1\right)$, so this is much less information than getting one digit of α. On the other hand, if the first continued fraction term is a 14, then $\alpha \in \left(\frac{1}{15}, \frac{1}{14}\right]$, so this pins down α to an interval of length $\frac{1}{15} - \frac{1}{14} = \frac{1}{210}$, which is much more information than a single decimal digit.

On average, if we learn one continued fraction term of α, how much information do we get, relative to the information we get from one decimal digit? If we get k decimal digits of α, then we have reduced the possible space for α by a factor of 10^k, so this suggests that the amount of information we get from a space reduction by a factor of b, relative to decimal digits, is $\log_{10} b$. Thus the information we get from learning that the first continued fraction term is 1 is $\log_{10} 2$, while the information we get from learning that the first continued fraction term is 14 is $\log_{10} 210$.

If we don't know what the first continued fraction term is, then we have to average over all possibilities, each weighted by their probability. Thus the information we expect to get from the first continued fraction term is

$$\sum_{n=1}^{\infty} \frac{1}{n(n+1)} \log_{10}(n(n+1)).$$

Remark 11.1. It is worth a warning that, while decimal digits are independent of each other, the same is not the case for continued fraction term, even with respect to Gauß measure! Check, for instance, that the (Gauß) measure of the set of $\alpha \in \mathbb{R}/\mathbb{Z}$ such that the first two continued fraction terms of α are both 1 is not the same as the product of the measures of the set of $\beta \in \mathbb{R}/\mathbb{Z}$ such that the first continued fraction term of β is 1, and the set of $\gamma \in \mathbb{R}/\mathbb{Z}$ such that the second continued fraction term of γ is 1. Thus we will have to work considerably harder to understand the average amount of information provided by a continued fraction term.

Of course, there's no particular reason to use decimal digits (hence base-10 logarithms) as our standard, so from now on we'll switch to the usual base-e logarithm.

Definition 11.2. Let X be a random variable taking on finitely many or countably many values, where each value x is taken on with probability $p(x)$. Then the *entropy* of X is

$$H(X) = -\sum_{x \in X} p(x) \log p(x).$$

Note that if $p(x) = 0$, then we encounter the undefined expression $0 \log 0$. Since $\lim_{x \to 0^+} x \log x = 0$, we treat $0 \log 0$ as 0 when it arises in entropy calculations. Also note that if X is a countable set, the sum $H(X)$ might diverge. For instance, this happens if $X = \mathbb{N}$ and $p(n) = \frac{1}{\sqrt{n}} - \frac{1}{\sqrt{n+1}}$. We will only be interested in situations where $H(X) < \infty$.

Now, suppose we get a string of information, one piece at a time. This could be decimal digits or continued fraction terms corresponding to some $\alpha \in [0, 1)$, or information about some other object presented in some other way. To formalize this, suppose $\mathcal{P}$ is an alphabet, i.e. a finite or countable set of "letters," each of which occurs with some probability; we'll write $p(a)$ for the probability of a letter a. Thus we have $p(a) \geq 0$ for all $a \in \mathcal{P}$, and $\sum_{a\in\mathcal{P}} p(a) = 1$. We then let $X = \mathcal{P}^{\mathbb{N}}$ be the set of all infinite sequences of elements of $\mathcal{P}$; we write x_n for the nth term of $x \in X$, so $x = (x_0, x_1, \ldots)$. We put a probability space structure on X by letting $\mathcal{A}$ be the σ-algebra generated by the cylinder sets

$$S_{b_0,\ldots,b_n} = \{x \in X : x_0 = b_0, x_1 = b_1, \ldots, x_n = b_n\}.$$

Then we let μ be the probability measure defined by

$$\mu(S_{b_0,\ldots,b_n}) = \prod_{i=0}^{n} p(b_i).$$

The entropy from one digit of X is

$$H(\mathcal{P}) = -\sum_{a\in\mathcal{P}} p(a)\log p(a).$$

The entropy from two digits of X is

$$-\sum_{a,b\in\mathcal{P}} p(a)p(b)\log p(a)p(b) = 2H(\mathcal{P}).$$

More generally, the entropy from knowing k digits of X is $kH(\mathcal{P})$. Thus knowledge of each additional x_n gives us an extra $H(\mathcal{P})$ information, so the information per term in the sequence is $H(\mathcal{P})$.

In more complicated situations, it might happen that not each piece of information provides exactly the same entropy. If we let H_n be the entropy from n pieces of information, then the average entropy per piece of information is $\frac{1}{n}H_n$. Then we define the entropy per piece of information to be

$$h = \lim_{n\to\infty} \frac{H_n}{n}.$$

Note that, in the case of X above, two digits of X are independent of each other, in the sense that knowing one of them provides no additional information about the other. In terms of probability, we have

$$\mathbb{P}(x_1 = b_1 \mid x_0 = b_0) = \mathbb{P}(x_1 = b_1) = p(b_1).$$

But sometimes we may be given two pieces of information that are dependent on each other. For example, knowledge of the first decimal digit of $\alpha \in [0, 1)$ provides at least partial information, and possibly full information, about the first continued fraction term, so if we know the first decimal digit, then the first continued fraction term is, on average, less useful than it would be if we didn't already know the first decimal digit.

Here is the general setup. Let $(X, \mathcal{A}, \mu)$ be a probability space, and let $\mathcal{P} = \{A_1, A_2, \ldots\}$ be a finite or countable partition of X, i.e. each $A_i \in \mathcal{A}$, and every $x \in X$ is contained in exactly one A_i, at least up to a null set. We call the A_i's the *atoms* of the partition. The entropy of the partition is

$$H(\mathcal{P}) = -\sum_n \mu(A_n) \log \mu(A_n).$$

Definition 11.3. Suppose $\mathcal{P} = \{A_1, A_2, \ldots\}$ and $\mathcal{Q} = \{B_1, B_2, \ldots\}$ are two partitions of X. Their *join* is

$$\mathcal{P} \vee \mathcal{Q} = \{A_i \cap B_j\}.$$

Of course, we are also free to take joins of more than two partitions. If $\mathcal{P}_1, \ldots, \mathcal{P}_k$ are partitions, then we write $\bigvee_{i=1}^k \mathcal{P}_i$ for the join of all k partitions.

Definition 11.4. With notation as above, let T be a measure-preserving transformation on $(X, \mathcal{A}, \mu)$. Then we define

$$T^{-1}\mathcal{P} = \{T^{-1}(A_1), T^{-1}(A_2), \ldots\}.$$

It is easy to check that if $\mathcal{P}$ and $\mathcal{Q}$ are partitions, then $\mathcal{P} \vee \mathcal{Q}$ and $T^{-1}\mathcal{P}$ are also partitions; you will do so in Problem 2.

Definition 11.5. Let $\mathcal{P}$ and $\mathcal{Q}$ be two partitions of $(X, \mathcal{A}, \mu)$. We say that $\mathcal{Q}$ is a *refinement* of $\mathcal{P}$, and write $\mathcal{P} \le \mathcal{Q}$, if for every $B \in \mathcal{Q}$, there is some $A \in \mathcal{P}$ such that $B \subseteq A$, up to a null set, i.e. $\mu(B \setminus A) = 0$.

Note that if $\mathcal{P} \le \mathcal{Q}$, then the atoms of $\mathcal{P}$ are larger than the atoms of $\mathcal{Q}$, but $\mathcal{Q}$ has more atoms, i.e. the atoms of $\mathcal{Q}$ are more informative than the atoms of $\mathcal{P}$. If $\mathcal{P} \le \mathcal{Q}$, then every atom in $\mathcal{P}$ is a union of atoms in $\mathcal{Q}$, up to a null set. Note that if $\mathcal{P}$ and $\mathcal{Q}$ are partitions of $(X, \mathcal{A}, \mu)$, then $\mathcal{P} \le \mathcal{P} \vee \mathcal{Q}$.

Here are some basic properties of the entropies of partitions:

Proposition 11.6. *Let $\mathcal{P}$ and $\mathcal{Q}$ be partitions of $(X, \mathcal{A}, \mu)$, and let T be a measure-preserving transformation. Then the following are true:*

- $H(T^{-1}\mathcal{P}) = H(\mathcal{P})$.
- *If* $\mathcal{P} \leq \mathcal{Q}$, *then* $H(\mathcal{P}) \leq H(\mathcal{Q})$.
- $H(\mathcal{P} \vee \mathcal{Q}) \leq H(\mathcal{P}) + H(\mathcal{Q})$.
- *We say that partitions $\mathcal{P}$ and $\mathcal{Q}$ are* independent *if for all $A \in \mathcal{P}$ and $B \in \mathcal{Q}$, we have $\mu(A \cap B) = \mu(A)\mu(B)$. If $\mathcal{P}$ and $\mathcal{Q}$ are independent, then $H(\mathcal{P} \vee \mathcal{Q}) = H(\mathcal{P}) + H(\mathcal{Q})$.*

You will prove Proposition 11.6 in Problem 3.

Sometimes, partitions are dependent, and giving information about one partition $\mathcal{Q}$ gives additional information about another partition $\mathcal{P}$. When this happens, we can define the *conditional entropy*.

Definition 11.7. Suppose $\mathcal{P} = \{A_1, A_2, \ldots\}$ and $\mathcal{Q} = \{B_1, B_2, \ldots\}$. Then the *conditional entropy* $H(\mathcal{P} \mid \mathcal{Q})$ is defined to be

$$H(\mathcal{P} \mid \mathcal{Q}) = -\sum_{A \in \mathcal{P}} \sum_{B \in \mathcal{Q}} \log \frac{\mu(A \cap B)}{\mu(B)} \cdot \mu(A \cap B).$$

If $\mathcal{P}$ and $\mathcal{Q}$ are independent, then $H(\mathcal{P} \mid \mathcal{Q}) = H(\mathcal{P})$, as you will verify in Problem 4.

Proposition 11.8. *Let $\mathcal{P}$, $\mathcal{Q}$, and $\mathcal{R}$ be three partitions of X.*

- $H(\mathcal{P} \vee \mathcal{Q} \mid \mathcal{R}) = H(\mathcal{P} \mid \mathcal{R}) + H(\mathcal{Q} \mid \mathcal{P} \vee \mathcal{R})$.
- *If* $\mathcal{Q} \leq \mathcal{R}$, *then* $H(\mathcal{P} \mid \mathcal{R}) \leq H(\mathcal{P} \mid \mathcal{Q})$.
- $H(\mathcal{P} \vee \mathcal{Q}) = H(\mathcal{P}) + H(\mathcal{Q} \mid \mathcal{P})$.
- *If* $\mathcal{P} \leq \mathcal{Q}$, *then* $H(\mathcal{P} \mid \mathcal{Q}) = 0$.

You will prove Proposition 11.8 in Problem 5.

Now let T be a measure-preserving transformation on $(X, \mathcal{A}, \mu)$, and let $\mathcal{P}$ be a partition. For a positive integer n, let $\mathcal{P}^{(n)}$ be the partition

$$\mathcal{P}^{(n)} = \bigvee_{i=0}^{n-1} T^{-i}\mathcal{P}.$$

Thus the atoms of $\mathcal{P}^{(n)}$ are of the form $A_{i_0} \cap T^{-1}(A_{i_1}) \cap \cdots \cap T^{-(n-1)}(A_{i_{n-1}})$ for some $i_0, \ldots, i_{n-1}$; it consists of those $x \in X$ such that $x \in A_{i_0}$, $T(x) \in A_{i_1}, \ldots, T^{n-1}(x) \in A_{i_{n-1}}$.

Definition 11.9. With notation as above, the *entropy* of T with respect to the partition $\mathcal{P}$ is

$$h(\mathcal{P}, T) = \lim_{n\to\infty} \frac{1}{n} H(\mathcal{P}^{(n)}).$$

The entropy of T is defined to be

$$h(T) = \sup_{\mathcal{P}} h(\mathcal{P}, T),$$

where the supremum is over all partitions of X.

Implicit in Definition 11.9 is that $\lim_{n\to\infty} \frac{1}{n} H(\mathcal{P}^{(n)})$ exists. The proof of this fact relies on a basic theorem from real analysis known as *Fekete's Lemma*.

Definition 11.10. Let $a_1, a_2, \ldots$ be a sequence in $\mathbb{R}$. We say that this sequence is *subadditive* if $a_{m+n} \le a_m + a_n$ for all $m, n \ge 1$.

Lemma 11.11 (Fekete, [**Fek23**]). *Let $a_1, a_2, \ldots$ be a subadditive sequence of nonnegative real numbers. Then*

$$\lim_{n\to\infty} \frac{a_n}{n} = \inf \frac{a_n}{n}.$$

Proof. Let $n, k \ge 1$, and let q, r be integers such that $n = kq + r$ and $0 \le r \le k - 1$. Then we have

$$\begin{aligned} \frac{a_n}{n} &= \frac{a_{kq+r}}{n} \\ &\le \frac{qa_k + a_r}{n} \\ &= \frac{qa_k}{kq+r} + \frac{a_r}{n} \\ &\le \frac{a_k}{k} + \frac{a_r}{n} \\ &\le \frac{a_k}{k} + \frac{\max(a_0, \ldots, a_{k-1})}{n}. \end{aligned}$$

As $n \to \infty$, the second term on the right tends to 0, so it follows that

$$\limsup_{n\to\infty} \frac{a_n}{n} \le \frac{a_k}{k}$$

for any fixed k, so

$$\limsup_{n\to\infty} \frac{a_n}{n} \le \inf_k \frac{a_k}{k}.$$

On the other hand, we have

$$\inf_k \frac{a_k}{k} \le \frac{a_n}{n},$$

so

$$\inf_k \frac{a_k}{k} \le \liminf_{n\to\infty} \frac{a_n}{n}.$$

Thus we have

$$\limsup_{n\to\infty} \frac{a_n}{n} \le \inf_k \frac{a_k}{k} \le \liminf_{n\to\infty} \frac{a_n}{n},$$

so both inequalities must be equalities, and $\lim_{n\to\infty} \frac{a_n}{n}$ exists and is equal to $\inf_k \frac{a_k}{k}$, as desired. ■

By Proposition 11.6, the sequence $H(\mathcal{P}^{(n)})$ is a subadditive sequence of nonnegative real numbers, so by Fekete's Lemma there exists

$$h(\mathcal{P}, T) = \lim_{n\to\infty} \frac{1}{n} H(\mathcal{P}^{(n)}).$$

Example 11.12. Let T be the transformation $T(x) = 10x \pmod 1$ on $(\mathbb{R}/\mathbb{Z}, \mathcal{L}, \lambda)$. Let $\mathcal{P}$ be the partition into 10 atoms $A_1, \dots, A_{10}$, where $A_i = \left[\frac{i-1}{10}, \frac{i}{10}\right)$. Then we have

$$H(\mathcal{P}) = -\sum_{n=1}^{10} \lambda(A_n) \log \lambda(A_n) = -\sum_{n=1}^{10} \frac{1}{10} \log \frac{1}{10} = \log 10.$$

Since all the $T^{-n}\mathcal{P}$'s are independent, we have $H(\mathcal{P}^{(n)}) = n \log 10$, so $h(\mathcal{P}, T) = \log 10$.

At this point, it is not clear whether there exists a better partition, i.e. one with a larger value of $h(\mathcal{P}, T)$. One possible attempt is to divide $\mathbb{R}/\mathbb{Z}$ into 100 evenly spaced intervals. But this does not increase the entropy; with this partition $\mathcal{Q}$, we would find that

$$H(\mathcal{Q}^{(n)}) = \begin{cases} (n+1)\log 10 & n \text{ odd}, \\ n \log 10 & n \text{ even}. \end{cases}$$

Thus we would get the same entropy as before from this partition.

Here are some basic properties of entropy:

Proposition 11.13. *Let $(X, \mathcal{A}, \mu, T)$ be a measure-preserving system. Let $\mathcal{P}$ and $\mathcal{Q}$ be partitions of X.*

- $h(\mathcal{P}, T) \leq H(\mathcal{P})$.
- $h(\mathcal{P} \vee \mathcal{Q}, T) \leq h(\mathcal{P}, T) + h(\mathcal{Q}, T)$.
- *If $\mathcal{P} \leq \mathcal{Q}$, then $h(\mathcal{P}, T) \leq h(\mathcal{Q}, T)$.*
- $h(\mathcal{P}, T) \leq h(\mathcal{Q}, T) + H(\mathcal{P} \mid \mathcal{Q})$.
- *For all $k \geq 1$, we have*
$$h(\mathcal{P}, T) = h(\mathcal{P}^{(k)}, T).$$
- *For all $k \geq 1$, we have $h(T^k) = k \cdot h(T)$.*

You will prove Proposition 11.13 in Problem 6.

Corollary 11.14. *If T is a rational rotation on $\mathbb{R}/\mathbb{Z}$, then $h(T) = 0$.*

Proof. Suppose $T = R_{\frac{p}{q}}$. Then T^q is the identity, which evidently has entropy 0. ∎

11.2 Computing Entropy with the Kolmogorov–Sinai Theorem

Based on the definition of entropy of a transformation, it seems that it is very hard to calculate it, because it requires looking at all partitions of the measure space. Might it be possible to select a partition $\mathcal{P}$ such that we can guarantee that $h(T) = h(\mathcal{P}, T)$?

Definition 11.15. Let $(X, \mathcal{A}, \mu)$ be a measure space, and let T be a measure-preserving transformation. Let $\mathcal{B}$ be a sub-σ-algebra of $\mathcal{A}$. We say that $\mathcal{B}$ is *T-invariant* if for all $B \in \mathcal{B}$, there exists some $C \in \mathcal{B}$ such that $\mu(T^{-1}(B) \triangle C) = 0$.

In other words, the preimage of every set in $\mathcal{B}$ is still in $\mathcal{B}$, up to a null set.

Definition 11.16. Let T be a measure-preserving transformation on $(X, \mathcal{A}, \mu)$. We say that a partition $\mathcal{P}$ of X is a *dynamical generator* with respect to T if the smallest T-invariant σ-algebra containing $\mathcal{P}$ is equal to $\mathcal{A}$, up to null sets.

In the case where X is a metric space and $\mathcal{A}$ is the Borel σ-algebra on X, then this says that the smallest T-invariant σ-algebra containing $\mathcal{P}$ is $\mathcal{A}$.

Example 11.17. Let K be a finite or countable set with a probability measure p on K, let $X = K^{\mathbb{N}}$ be the set of sequences with probability measure μ defined such that if $S_{b_0,b_1,\ldots,b_n}$ is a cylinder set, then we have

$$\mu(S_{b_0,\ldots,b_n}) = \prod_{i=0}^{n} p(b_i),$$

as usual. Let $\mathcal{A}$ be the σ-algebra on X generated by the cylinder sets. Then the collection of cylinder sets $\{S_k : k \in K\}$ forms a dynamical generator for the Bernoulli shift T on X. Here,

$$S_k = \{(x_0, x_1, \ldots) \in X : x_0 = k\},$$

and T is defined by

$$T(x_0, x_1, \ldots) = (x_1, x_2, \ldots).$$

Theorem 11.18 (Kolmogorov–Sinai). *Let T be a measure-preserving transformation on $(X, \mathcal{A}, \mu)$. If a partition $\mathcal{P}$ is a dynamical generator for T, then $h(T) = h(\mathcal{P}, T)$.*

The Kolmogorov–Sinai Theorem makes it much easier to calculate the entropy of a transformation T, because now we only have to find one dynamical generator for T and calculate its entropy, rather than considering all possible partitions.

For instance, the Kolmogorov–Sinai Theorem immediately tells us how to compute the entropy of a Bernoulli shift. Let T be a Bernoulli shift and K the partition given by looking at the first element in the sequence. Then K is a dynamical generator for T. The $T^{-n}(K)$'s are independent of each other, for different values of n. Thus

$$h(K, T) = H(K) = -\sum_{k \in K} p(k) \log p(k).$$

Let's now prove the Kolmogorov–Sinai Theorem. We begin with a couple lemmas. The first lemma says that if we have two finite partitions $\mathcal{P} = \{P_1, P_2, \dots, P_r\}$ and $\mathcal{Q} = \{Q_1, Q_2, \dots, Q_r\}$ that are close together in the sense that $\sum_{i=1}^r \mu(P_i \vartriangle Q_i)$ is small, then their conditional entropies $H(\mathcal{P} \mid \mathcal{Q})$ and $H(\mathcal{Q} \mid \mathcal{P})$ are also small.

Lemma 11.19. *Let r be a positive integer. For every $\varepsilon > 0$, there exists a $\delta > 0$ such that if $\mathcal{P} = \{P_1, \dots, P_r\}$ and $\mathcal{Q} = \{Q_1, \dots, Q_r\}$ are two partitions of $(X, \mathcal{A}, \mu)$ into r atoms, and $\sum_{i=1}^r \mu(P_i \vartriangle Q_i) < \delta$, then $H(\mathcal{P} \mid \mathcal{Q}) + H(\mathcal{Q} \mid \mathcal{P}) < \varepsilon$.*

Proof. Fix $\varepsilon > 0$, and choose $\delta > 0$ such that $\delta < \frac{1}{4}$ and

$$-r(r-1)\delta \log \delta - (1-\delta)\log(1-\delta) < \frac{\varepsilon}{2}.$$

Let $\mathcal{R}$ be the partition of X into sets $P_i \cap Q_j$ for $i \neq j$, together with $\bigcup_{i=1}^r (P_i \cap Q_j)$. Then $\mathcal{P} \vee \mathcal{Q} = \mathcal{P} \vee \mathcal{R} = \mathcal{Q} \vee \mathcal{R}$, and since if $i \neq j$ we have $P_i \cap Q_j \subseteq \bigcup_{n=1}^r (P_n \vartriangle Q_n)$, we have $\mu(P_i \cap Q_j) < \delta$ for $i \neq j$ and

$$\mu\left(\bigcup_{i=1}^r (P_i \cap Q_i)\right) = \sum_{i=1}^r \mu(P_i \cap Q_i) > 1 - \delta.$$

Thus we have

$$H(\mathcal{R}) < r(r-1)\delta \log \delta - (1-\delta)\log(1-\delta) < \frac{\varepsilon}{2}.$$

Thus

$$\begin{aligned}
H(\mathcal{Q}) + H(\mathcal{P} \mid \mathcal{Q}) &= H(\mathcal{P} \vee \mathcal{Q}) \\
&= H(\mathcal{Q} \vee \mathcal{R}) \\
&\leq H(\mathcal{Q}) + H(\mathcal{R}) \\
&< H(\mathcal{Q}) + \frac{\varepsilon}{2},
\end{aligned}$$

so $H(\mathcal{P} \mid \mathcal{Q}) < \frac{\varepsilon}{2}$. Similarly, $H(\mathcal{Q} \mid \mathcal{P}) < \frac{\varepsilon}{2}$, and the result follows. ■

We now need a similar result to bound the conditional entropy of a countable partition by that of a related finite partition. Given a countable partition, we can turn it into a finite partition, say with $k+1$ atoms, by merging atoms $k+1, k+2, \dots$ into a single atom.

Lemma 11.20. *Let $\mathcal{P} = \{P_1, P_2, \ldots\}$ be a countable partition with finite entropy, and let*

$$\mathcal{P}_k = \left\{P_1, \ldots, P_k, X \setminus \bigcup_{i=1}^{k} P_i\right\}.$$

Then $\lim_{k\to\infty} H(\mathcal{P} \mid \mathcal{P}_k) = 0$.

Proof. Let $P_0 = X \setminus \bigcup_{i=1}^{k} P_i$. Then we have

$$H(\mathcal{P} \mid \mathcal{P}_k) = \sum_{i=0}^{k} \sum_{j=1}^{\infty} -\mu(P_i \cap P_j) \log \frac{\mu(P_i \cap P_j)}{\mu(P_i)}.$$

For any term with $i \geq 1$ and $j \neq i$, we have $\mu(P_i \cap P_j) = 0$, so those terms vanish. On the other hand, if $j = i$, then we have $\frac{\mu(P_i \cap P_j)}{\mu(P_i)} = 1$, so its logarithm is zero, and those terms also vanish. If $i = 0$, then we have $\mu(P_i \cap P_j) = 0$ if $j \leq k$ and $\mu(P_i \cap P_j) = \mu(P_j)$ if $j > k$. Thus we have

$$H(\mathcal{P} \mid \mathcal{P}_k) = \sum_{j=k+1}^{\infty} -\mu(P_j) \log \frac{\mu(P_j)}{\mu(P_0)} \leq \sum_{j=k+1}^{\infty} -\mu(P_j) \log \mu(P_j).$$

Since we are assuming that $\mathcal{P}$ has finite entropy, the last sum tends to 0 as $k \to \infty$. ■

We now show that, given a generating partition $\mathcal{Q}$, we can approximate any partition arbitrarily well by a finite subpartition of $\mathcal{Q}$.

Proposition 11.21. *Let $\mathcal{Q}$ be a partition of X that generates the σ-algebra $\mathcal{A}$ up to null sets. For every partition $\mathcal{P}$ of X with $H(\mathcal{P}) < \infty$, there exists a finite partition $\mathcal{R}$ of X with $\mathcal{R} \leq \mathcal{Q}$ such that $H(\mathcal{P} \mid \mathcal{R}) < \varepsilon$.*

Proof. Fix $\varepsilon > 0$, and find a k such that $H(\mathcal{Q} \mid \mathcal{Q}_k) < \frac{\varepsilon}{2}$, using the notation from Lemma 11.20. Because $\mathcal{Q}$ generates $\mathcal{A}$ up to null sets, for each i we can find some A_i which is a finite union of atoms of $\mathcal{Q}$ such that

$$\mu(P_i \vartriangle A_i) < \frac{\delta}{2k^2}.$$

Let $Q_1 = A_1$ and $Q_i = A_i \setminus \bigcup_{j=1}^{i-1} A_j$ for $2 \leq i \leq k$, and let $Q_0 = X \setminus \bigcup_{j=1}^{k-1} A_j$. Then $\mathcal{Q}_k = \{Q_0, Q_1, \ldots, Q_k\}$ is a partition of X, and each Q_i

is a finite union of atoms in $\mathcal{Q}$. For $1 \leq i \leq k$, we have

$$Q_i \vartriangle A_i = Q_i \setminus A_i = A_i \cap \bigcup_{j=1}^{i-1} A_j.$$

For every $x \in Q_i \vartriangle A_i$, there is some $j < i$ such that $x \in A_i \cap A_j$. Since $P_i \cap P_j = \varnothing$, we have $x \in (A_i \setminus P_i) \cap (A_j \setminus P_j)$. Thus we have

$$Q_i \vartriangle A_i \subseteq \bigcup_{j=1}^{i} (A_j \setminus P_j) \subseteq \bigcup_{j=1}^{i} (A_j \vartriangle P_j),$$

so

$$\mu(Q_i \vartriangle A_i) < \frac{i\delta}{2k^2} \leq \frac{\delta}{2k}.$$

Thus we have

$$\mu(P_i \vartriangle Q_i) \leq \mu(P_i \vartriangle A_i) + \mu(Q_i \vartriangle A_i) < \frac{\delta}{2k^2} + \frac{\delta}{2k} \leq \frac{\delta}{k}.$$

Next, we have $P_0 \vartriangle Q_0 \subseteq \bigcup_{i=1}^{k} P_i \vartriangle Q_i$, so $\mu(P_0 \vartriangle Q_0) < \delta$. By Lemma 11.19, if δ is sufficiently small, this implies that $H(\mathcal{P} \mid \mathcal{P}_k)$ and $H(\mathcal{P}_k \mid \mathcal{Q})$ are both $< \frac{\varepsilon}{2}$, so we have

$$H(\mathcal{P} \mid \mathcal{Q}) \leq H(\mathcal{P} \mid \mathcal{P}_k) + H(\mathcal{P}_k \mid \mathcal{Q}) < \varepsilon,$$

as desired. ■

Corollary 11.22. *If $\mathcal{P}$ is a dynamical generator for a transformation T and $\mathcal{Q}$ is a partition with finite entropy, then*

$$\lim_{n\to\infty} H(\mathcal{Q} \mid \mathcal{P}^{(n)}) = 0.$$

Proof. Let $\mathcal{A}_n$ be the σ-algebra generated by $\mathcal{P}^{(n)}$. Then, because $\mathcal{P}$ is a dynamical generator, the σ-algebra generated by $\bigcup_{n=1}^{\infty} \mathcal{A}_n$ is $\mathcal{A}$, up to null sets. Let $\varepsilon > 0$ be arbitrary. Then by Proposition 11.21, there exists a finite partition $\mathcal{R} \subseteq \mathcal{A}$ such that $H(\mathcal{Q} \mid \mathcal{R}) < \varepsilon$. Since $\mathcal{R}$ is finite, there exists some m such that $\mathcal{R} \leq \mathcal{P}^{(m)}$. Thus for all $n \geq m$, we have

$$H(\mathcal{Q} \mid \mathcal{P}^{(n)}) \leq H(\mathcal{Q} \mid \mathcal{P}^{(m)}) \leq H(\mathcal{Q} \mid \mathcal{R}) < \varepsilon,$$

as desired. ■

We are now ready to complete the proof of the Kolmogorov–Sinai Theorem.

Proof of the Kolmogorov–Sinai Theorem. It suffices to show that if $\mathcal{Q}$ is any partition with finite entropy, then

$$h(\mathcal{Q}, T) \leq h(\mathcal{P}, T).$$

We have

$$h(\mathcal{Q}, T) \leq h(\mathcal{P}^{(n)}, T) + H(\mathcal{Q} \mid \mathcal{P}^{(n)})$$

for all n. By Corollary 11.22, we have $\lim_{n\to\infty} H(\mathcal{Q} \mid \mathcal{P}^{(n)}) = 0$, so we have

$$h(\mathcal{Q}, T) \leq \lim_{n\to\infty} H(\mathcal{P}^{(n)}) = h(\mathcal{P}, T),$$

as desired. ■

The next theorem guarantees that, under reasonable circumstances, finite dynamical generators exist.

Theorem 11.23 (Krieger, [**Kri70**]). *If T is an invertible ergodic transformation with $h(T) < \infty$, then there is a finite dynamical generator for T.*

11.3 Conditional Expectation with Measure Theory

Our next major result will be the Shannon–McMillan–Breiman Theorem, which gives a local interpretation of entropy as a time average, analogous to the Birkhoff Ergodic Theorem. But before we can get there, we need some background on expected values from the measure-theoretic perspective. In classical probability, without measure theory, if X is a discrete random variable on a sample space Ω and A is an event, then the conditional expectation $\mathbb{E}(X \mid A)$ is defined as

$$\mathbb{E}(X \mid A) = \sum_{\omega\in\Omega} X(\omega)\mathbb{P}(\omega \mid A).$$

Or, if we have a continuous random variable X on a probability space $(\Omega, \mathcal{A}, \mu)$ and $A \in \mathcal{A}$, we have

$$\mathbb{E}(X \mid A) = \frac{1}{\mu(A)} \int_A X \, d\mu.$$

To make further progress, we need a definition of conditional expectation that allows us to condition not just on a measurable subset, but on a smaller σ-algebra, and here the definition is a little more complicated.

Definition 11.24. A *random variable* on a probability space $(\Omega, \mathcal{A}, \mu)$ is a measurable function $X : \Omega \to \mathbb{R}$.

Definition 11.25. Let $(\Omega, \mathcal{A}, \mu)$ be a probability space, X a random variable on Ω, and $\mathcal{B}$ a sub-σ-algebra of $\mathcal{A}$. The *conditional expectation* $\mathbb{E}(X \mid \mathcal{B})$ is a $\mathcal{B}$-measurable function such that

$$\int_B \mathbb{E}(X \mid \mathcal{B})\, d\mu = \int_B X\, d\mu$$

for all $B \in \mathcal{B}$.

Note that we cannot typically take $\mathbb{E}(X \mid \mathcal{B}) = X$, because X need not be $\mathcal{B}$-measurable.

Based on the definition, it might not be clear that a function $\mathbb{E}(X \mid \mathcal{B})$ necessarily exists, or is unique (up to a null set), but these both turn out to be true. See [**CK04**, §5.4.3] for a proof.

11.4 The Shannon–McMillan–Breiman Theorem

Let $(X, \mathcal{A}, \mu, T)$ be a measure-preserving system, and let $\mathcal{P}$ be a partition of X. As we have seen, for each n, $\mathcal{P}^{(n)} = \bigvee_{i=0}^{n-1} T^{-i}\mathcal{P}$ is again a partition of X. If we start with some $x \in X$, we can look at which atom in $\mathcal{P}^{(n)}$ x lies in. Write $A_n(x)$ for the atom in $\mathcal{P}^{(n)}$ containing x. What should we expect $\mu(A_n(x))$ to be? This is the content of the Shannon–McMillan–Breiman Theorem.

Theorem 11.26 (Shannon, McMillan, and Breiman, [**Sha48**], [**McM53**], [**Bre57**]). *With notation as above, suppose that $H(\mathcal{P}) < \infty$. Then for almost all $x \in X$, we have*

$$\lim_{n\to\infty} -\frac{1}{n} \log \mu(A_n(x)) = h(\mathcal{P}, T).$$

We need to set up a few ingredients before turning to the proof of the Shannon–McMillan–Breiman Theorem.

Definition 11.27. Let $\mathcal{P}$ be a partition of a measure space $(X, \mathcal{A}, \mu)$. The *information function* associated to $\mathcal{P}$ is the function $I_{\mathcal{P}} : X \to \mathbb{R}_{\geq 0} \cup \{\infty\}$

defined by

$$I_{\mathcal{P}}(x) = -\log \mu(P(x)),$$

where $P(x)$ is the atom of $\mathcal{P}$ containing x.

The information function is closely related to the entropy: we have

$$H(\mathcal{P}) = \int_X I_{\mathcal{P}}\, d\mu.$$

The information function is local, depending on a point $x \in X$, whereas the entropy is the global, averaged, version.

We also have a conditional version of the information function: if $\mathcal{P}$ and $\mathcal{Q}$ are two partitions of X, then the conditional information of $\mathcal{P}$ given $\mathcal{Q}$ is defined to be

$$I_{\mathcal{P}|\mathcal{Q}}(x) = -\sum_{P \in \mathcal{P}} \sum_{Q \in \mathcal{Q}} \mathbb{1}_{P \cap Q}(x) \log \frac{\mu(P \cap Q)}{\mu(Q)}.$$

We can also write $I_{\mathcal{P}|\mathcal{Q}}(x)$ as

$$I_{\mathcal{P}|\mathcal{Q}}(x) = -\log \mathbb{E}(\mathbb{1}_{P(x)} \mid \sigma(\mathcal{Q})) = -\sum_{P \in \mathcal{P}} \mathbb{1}_P(x) \log \mathbb{E}(\mathbb{1}_P \mid \sigma(\mathcal{Q})).$$

Here $\sigma(\mathcal{Q})$ denotes the σ-algebra generated by $\mathcal{Q}$, i.e. the smallest σ-algebra containing $\mathcal{Q}$. Clearly, we have

$$H(\mathcal{P} \mid \mathcal{Q}) = \int_X I_{\mathcal{P}|\mathcal{Q}}\, d\mu.$$

If $\mathcal{P}$ and $\mathcal{Q}$ are partitions of X, then

$$I_{\mathcal{P} \vee \mathcal{Q}} = I_{\mathcal{Q}} + I_{\mathcal{P}|\mathcal{Q}}.$$

Now, suppose T is measure-preserving on $(X, \mathcal{A}, \mu)$. If $\mathcal{P} = \{P_1, P_2, \ldots\}$ is a partition of X, then we have

$$\begin{aligned} I_{T^{-1}\mathcal{P}}(x) &= -\sum_{P \in \mathcal{P}} \mathbb{1}_{T^{-1}(P)}(x) \log \mu(T^{-1}(P)) \\ &= -\sum_{P \in \mathcal{P}} \mathbb{1}_P(T(x)) \log \mu(P) \\ &= I_{\mathcal{P}}(T(x)). \end{aligned}$$

Furthermore, we have

$$\lim_{n\to\infty} \frac{1}{n} H(\mathcal{P}^{(n)}) = \lim_{n\to\infty} \frac{1}{n} \int_X I_{\mathcal{P}^{(n)}}\, d\mu = h(\mathcal{P}, T).$$

Next, we need a key theorem from probability, which is a special case of the Martingale Convergence Theorem.

Theorem 11.28 (Martingale Convergence Theorem). *Let $\mathcal{A}_1 \subseteq \mathcal{A}_2 \subseteq \cdots$ be an increasing sequence of σ-algebras on X, and let $\mathcal{A}$ be the σ-algebra generated by $\bigcup_{n=1}^{\infty} \mathcal{A}_n$. If μ is a probability measure on $(X, \mathcal{A})$ and $f \in L^1(\mu)$, then*

$$\mathbb{E}_\mu(f \mid \mathcal{A}) = \lim_{n\to\infty} \mathbb{E}_\mu(f \mid \mathcal{A}_n)$$

almost everywhere, and $\mathbb{E}_\mu(f \mid \mathcal{A}) \in L^1(\mu)$.

See [**Wil91**, Theorem 11.5, p. 109] for a somewhat more general statement and a proof.

It follows that

$$I_{\mathcal{P}|\bigvee_{n=1}^{\infty} T^{-n}\mathcal{P}}(x) = \lim_{n\to\infty} I_{\mathcal{P}|\bigvee_{i=1}^{n} T^{-i}\mathcal{P}}(x)$$

for almost all $x \in X$.

We're now ready to prove the Shannon–McMillan–Breiman Theorem. Note that the statement of the theorem can be rewritten in terms of the information function: it states that

$$\lim_{n\to\infty} \frac{1}{n} I_{\mathcal{P}(n)}(x) = h(\mathcal{P}, T)$$

for almost all $x \in X$.

Proof of the Shannon–McMillan–Breiman Theorem. For a positive integer n, define

$$f_n(x) = I_{\mathcal{P}|\bigvee_{i=1}^{n-1} T^{-i}\mathcal{P}}(x).$$

We claim that for each positive integer n, we have

$$I_{\mathcal{P}(n)}(x) = I_{\mathcal{P}}(T^{n-1}(x)) + \sum_{i=2}^{n} f_i(T^{n-i}(x)).$$

We prove this by induction on n, with the case $n = 1$ saying that

$$I_{\mathcal{P}}(x) = I_{\mathcal{P}}(x),$$

which is clearly true. Assuming the result is true for $n-1$, we have

$$\begin{aligned}
I_{\mathcal{P}(n)}(x) &= I_{\bigvee_{i=1}^{n-1} T^{-i}\mathcal{P}}(x) + I_{\mathcal{P}|\bigvee_{i=1}^{n-1} T^{-i}\mathcal{P}}(x) \\
&= I_{\mathcal{P}(n-1)}(T(x)) + f_n(x) \\
&= I_{\mathcal{P}}(T^{n-2}(T(x))) + \sum_{i=2}^{n-1} f_i(T^{n-1-i}(T(x))) + f_n(x) \\
&= I_{\mathcal{P}}(T^{n-1}(x)) + \sum_{i=2}^{n} f_i(T^{n-i}(x)),
\end{aligned}$$

as claimed.

Now let

$$f(x) = I_{\mathcal{P}|\bigvee_{i=1}^{\infty} T^{-i}\mathcal{P}}(x) = \lim_{n\to\infty} f_n(x)$$

by the result following the Martingale Convergence Theorem. We have

$$\int_X f\,d\mu = h(\mathcal{P}, T),$$

so $f \in L^1(\mu)$. Let $f_1 = I_{\mathcal{P}}$. Then we have

$$\begin{aligned}
\frac{1}{n} I_{\mathcal{P}(n)}(x) &= \frac{1}{n}\sum_{i=0}^{n-1} f_{n-i}(T^i(x)) \\
&= \frac{1}{n}\sum_{i=0}^{n-1} f(T^i(x)) + \frac{1}{n}\sum_{i=0}^{n-1} (f_{n-i} - f)(T^i(x)).
\end{aligned}$$

By the Birkhoff Ergodic Theorem, we have

$$\lim_{n\to\infty} \frac{1}{n}\sum_{i=0}^{n-1} f(T^i(x)) = \int_X f\,d\mu = h(\mathcal{P}, T)$$

for almost all $x \in X$, so it suffices to show that the second sum tends to 0 for almost all $x \in X$.

Let

$$F_N(x) = \sup_{k\geq N} |f_k(x) - f(x)|, \qquad f^*(x) = \sup_{n\geq 1} f_n(x).$$

We have $0 \leq F_N \leq f^* + f$, and $f^* \in L^1(\mu)$ (see [**Daj14**, Proposition 5.4.1, pp. 92–94] for a proof), so $F_N \in L^1(\mu)$. Note also that $\lim_{N\to\infty} F_N(x) = 0$

for almost all $x \in X$. Thus by the Dominated Convergence Theorem, we have

$$\lim_{N\to\infty} \int_X F_N \, d\mu = 0.$$

For all i, we have $|f_{n-i}(x) - f(x)| \le f^*(x) + f(x)$, so $|f_{n-i} - f| \in L^1(\mu)$ and $\lim_{n\to\infty} |f_{n-i}(x) - f(x)| = 0$ for almost all $x \in X$.

For all $N \ge 1$ and all $n > N$, we have

$$\begin{aligned}\frac{1}{n}\sum_{i=0}^{n-1} |f_{n-i} - f|(T^i(x)) &= \frac{1}{n}\sum_{i=0}^{n-1-N} |f_{n-i} - f|(T^i(x)) \\ &\quad + \frac{1}{n}\sum_{i=n-N}^{n} |f_{n-i} - f|(T^i(x)) \\ &\le \frac{1}{n}\sum_{i=0}^{n-1-N} F_N(T^i(x)) \\ &\quad + \frac{1}{n}\sum_{i=0}^{N} |f_i - f|(T^{n-i}(x)).\end{aligned}$$

By Problem 5 in Chapter 4, the second term tends to 0 as $N \to \infty$ almost everywhere, and by the Birkhoff Ergodic Theorem, the first term tends to $\int_X F_N \, d\mu = 0$ almost everywhere. Thus we have

$$\lim_{n\to\infty} \frac{1}{n}\sum_{i=0}^{n-1} |f_{n-i} - f|(T^i(x)) = 0$$

almost everywhere. This completes the proof. ■

11.5 Lochs's Theorem

We now turn to the question we asked at the beginning: how much information does a continued fraction term give compared to a decimal digit? Lochs [**Loc63**] explored this question by determining the number of continued fraction terms that can be calculated from the first 1000 decimal digits of π, and he found that it is possible to determine 968 continued fraction terms. Similarly, Brent and McMillan [**BM80**] showed that 29200 continued fraction terms of the Euler–Mascheroni constant can be computed from the first 30100 decimal digits. This suggests that the continued fraction is a slightly more efficient encoding than the decimal expansion.

Let $\alpha \in [0, 1)$ have decimal expansion $.d_1 d_2 \ldots$ and continued fraction expansion $\alpha = [0; a_1, a_2, \ldots]$. Let y be the number obtained from the first n digits of the decimal expansion of α, i.e. $y = .d_1 d_2 \ldots d_n$, and let $[0; c_1, c_2, \ldots, c_k]$ be the continued fraction expansion of y. Let $m(n, \alpha)$ be the largest nonnegative integer m such that $a_i = c_i$ for all $i \leq m$, i.e.

$$a_1 = c_1, a_2 = c_2, \ldots, a_m = c_m, a_{m+1} \neq c_{m+1}.$$

Theorem 11.29 (Lochs, [**Loc64**]). *For almost all $\alpha \in [0, 1)$, we have*

$$\lim_{n\to\infty} \frac{m(n, \alpha)}{n} = \frac{6 \log 2 \cdot \log 10}{\pi^2} \approx 0.97027014\ldots$$

In order to prove Lochs's Theorem, we want to make use of the Shannon–McMillan–Breiman Theorem. But we need a bit of care. We need to compare entropies of two transformations on $\mathbb{R}/\mathbb{Z}$: the decimal expansion transformation $x \mapsto 10x - \lfloor 10x \rfloor$ and the continued fraction transformation $x \mapsto \frac{1}{x} - \left\lfloor \frac{1}{x} \right\rfloor$. But they are not measure-preserving with respect to the same measure: we need to use Lebesgue measure λ for the decimal expansion transformation and Gauß measure μ for the continued fraction transformation. So the first thing we need to do is to estimate how these relate to each other.

Lemma 11.30. *For any $A \in \mathcal{L}$, we have*

$$\frac{1}{2\log 2}\lambda(A) \leq \mu(A) \leq \frac{1}{\log 2}\lambda(A).$$

Proof. For any $A \in \mathcal{L}$, we have

$$\mu(A) = \frac{1}{\log 2} \int_A \frac{d\lambda}{1 + x}.$$

Since $\frac{1}{2} \leq \frac{1}{1+x} \leq 1$ on $[0, 1)$, we have the stated inequalities. ■

We must now look at the Lebesgue measures of cylinder sets for the continued fraction transformation T. For $x \in [0, 1)$, let $\Delta_n(x)$ be the nth cylinder set of x with respect to T, i.e. the set of numbers in $[0, 1)$ with the same first n terms in the continued fraction as x. Since the measure (either Lebesgue or Gauß) of the $\Delta_n(x)$'s tend to 0, Lemma 11.30 implies that for all irrational x, we have

$$\lim_{n\to\infty} \frac{\log \lambda(\Delta_n(x))}{\log \mu(\Delta_n(x))} = 1.$$

The next order of business is to estimate the measure of $\Delta_n(x)$. To do this, we need some more results on continued fractions.

Theorem 11.31 (Lévy, [**Lév29**]).

(1) *For almost all $x \in [0,1)$, we have*

$$\lim_{n\to\infty} \frac{1}{n} \log \lambda(\Delta_n(x)) = -\frac{\pi^2}{6\log 2}.$$

(2) *If we let the nth convergent of the continued fraction of x be $\frac{p_n(x)}{q_n(x)}$, then for almost all x, we have*

$$\lim_{n\to\infty} \frac{1}{n} \log q_n(x) = \frac{\pi^2}{12\log 2}.$$

(3) *For almost all $x \in [0,1)$, we have*

$$\lim_{n\to\infty} \frac{1}{n} \log\left| x - \frac{p_n(x)}{q_n(x)} \right| = -\frac{\pi^2}{6\log 2}.$$

Proof. We have

$$\begin{aligned}
\frac{p_n(x)}{q_n(x)} &= \frac{1}{a_1 + [a_2, a_3, \dots, a_n]} \\
&= \frac{1}{a_1 + \frac{p_{n-1}(T(x))}{q_{n-1}(T(x))}} \\
&= \frac{q_{n-1}(T(x))}{p_{n-1}(T(x)) + q_{n-1}(T(x))a_1}.
\end{aligned}$$

Thus we have $p_n(x) = q_{n-1}(T(x))$, and we have

$$\frac{1}{q_n(x)} = \frac{p_n(x)}{q_n(x)} \cdot \frac{p_{n-1}(T(x))}{q_{n-1}(T(x))} \cdots \frac{p_1(T^{n-1}(x))}{q_1(T^{n-1}(x))},$$

so we have

$$-\log q_n(x) = \sum_{i=0}^{n-1} \log \frac{p_{n-i}(T^i(x))}{q_{n-1}(T^i(x))}.$$

Since $\frac{p_k(x)}{q_k(x)}$ is close to x, this is close to what we get when we replace all the convergents of x with x itself, so let us write

$$-\log q_n(x) = \sum_{i=0}^{n-1} \log T^i(x) + R_n(x),$$

where

$$R_n(x) = \sum_{i=0}^{n-1} \left[\log T^i(x) - \log \frac{p_{n-i}(T^i(x))}{q_{n-i}(T^i(x))} \right].$$

Now, if we look at the average of the first term, by the Birkhoff Ergodic Theorem, we have

$$\lim_{n\to\infty} \frac{1}{n} \sum_{i=0}^{n-1} \log T^i(x) = \int_{\mathbb{R}/\mathbb{Z}} \log x \, d\lambda = \frac{1}{\log 2} \int_0^1 \frac{\log x}{1+x} \, dx.$$

We now evaluate this integral:

$$\begin{aligned}
\frac{1}{\log 2} \int_0^1 \frac{\log x}{1+x} \, dx &= \frac{1}{\log 2} \int_0^1 \log x \sum_{n=0}^{\infty} (-1)^n x^n \, dx \\
&= \frac{1}{\log 2} \sum_{n=0}^{\infty} (-1)^n \int_0^1 x^n \log x \, dx \\
&= \frac{1}{\log 2} \sum_{n=0}^{\infty} (-1)^n \left[\frac{x^{n+1}}{n+1} \log x \Big|_0^1 - \int_0^1 \frac{x^n}{n+1} \, dx \right] \\
&= -\frac{1}{\log 2} \sum_{n=0}^{\infty} \frac{(-1)^n}{(n+1)^2} \\
&= -\frac{\pi^2}{12 \log 2}.
\end{aligned}$$

So, we must now show that $\log_{n\to\infty} \frac{1}{n} R_n(x) = 0$. Note that if $x = [a_1, a_2, \ldots]$, then x lies between $[a_1, a_2, \ldots, a_n]$ and $[a_1, a_2, \ldots, a_n + 1]$, which defines the cylinder set Δ_n, so Δ_n has endpoints $[a_1, \ldots, a_n] = \frac{p_n(x)}{q_n(x)}$ and $[a_1, \ldots, a_n + 1] = \frac{p_n(x)+p_{n-1}(x)}{q_n(x)+q_{n-1}(x)}$. If n is even, then $x > \frac{p_n(x)}{q_n(x)}$, so by the Mean Value Theorem, there is some $\xi \in \left(\frac{p_n(x)}{q_n(x)}, x\right)$ such that

$$\begin{aligned} 0 &< \log x - \log \frac{p_n(x)}{q_n(x)} \\ &= \left(x - \frac{p_n(x)}{q_n(x)}\right)\frac{1}{\xi} \\ &\leq \frac{1}{q_n(x)(q_{n-1}(x)+q_n(x))} \cdot \frac{1}{p_n(x)/q_n(x)} \\ &< \frac{1}{q_n(x)}. \end{aligned}$$

Similarly,

$$\frac{1}{q_n(x)} < \log x - \log \frac{p_n(x)}{q_n(x)}$$

when n is odd. Let F_n be the nth Fibonacci number. Then we have $q_n(x) \geq F_n$. Thus we have

$$|R_n(x)| \leq \sum_{i=1}^{n} \frac{1}{F_i}.$$

Since $F_n \approx \frac{1}{\sqrt{5}}\left(\frac{1+\sqrt{5}}{2}\right)^n$, the sum of the reciprocals of the Fibonacci numbers converges to some number ψ. Thus we have $|R_n(x)| \leq \psi$ for all n and x, which implies that $\lim_{n\to\infty} \frac{1}{n}R_n(x) = 0$. This completes the proof of (2).

To get (1), note that

$$\lambda(\Delta_n(a_1, \ldots, a_n)) = \frac{1}{q_n(x)(q_n(x)+q_{n+1}(x))}.$$

Thus we have

$$-\log 2 - 2\log q_n(x) < \log \lambda(\Delta_n) < -2\log q_n(x).$$

Then part (2) implies part (1).

To get part (3), we use the fact that

$$\frac{1}{2q_n(x)q_{n+1}(x)} < \left|x - \frac{p_n(x)}{q_n(x)}\right| < \frac{1}{q_n(x)q_{n+1}(x)};$$

see [**Khi64**, Theorem 9, p. 9 and Theorem 13, p. 15]. The result then follows from part (2). ■

Let's now use the Kolmogorov–Sinai Theorem and the Shannon–McMillan–Breiman Theorem to compute the entropy $h(T)$ of the continued fraction transformation T on $\mathbb{R}/\mathbb{Z}$, given by $T(x) = \frac{1}{x} - \left\lfloor \frac{1}{x} \right\rfloor$. Because these theorems require that T be measure-preserving, we need to use the Gauß measure μ and the Lebesgue σ-algebra $\mathcal{L}$. One can check that the partition $\mathcal{P}$ given by $A_n = \left(\frac{1}{n}, \frac{1}{n+1}\right]$ is a dynamical generator for T. Thus by the Kolmogorov–Sinai Theorem, it suffices to compute $h(\mathcal{P}, T)$. Now, the Shannon–McMillan–Breiman Theorem tells us that

$$\lim_{n\to\infty} -\frac{1}{n} \log \mu(\Delta_n(x)) = h(\mathcal{P}, T).$$

And since

$$\lim_{n\to\infty} \frac{\log \lambda(\Delta_n(x))}{\log \mu(\Delta_n(x))} = 1,$$

it suffices to compute

$$\lim_{n\to\infty} -\frac{1}{n} \log \lambda(\Delta_n(x)).$$

But as we have seen, this is $\frac{\pi^2}{6\log 2}$.

There is another way to compute $h(T)$, using *Rokhlin's formula*. See Problems 13 and 14.

Let's write $\Delta(a_1, \dots, a_n)$ for the cylinder set under T of $a_1, \dots, a_n$, i.e.

$$\Delta(a_1, \dots, a_n) = \left\{ x \in [0,1) \,:\, T^{i-1}(x) \in \left(\frac{1}{a_i}, \frac{1}{a_i + 1}\right] \text{ for } 1 \le i \le n \right\}.$$

Given a cylinder set $\Delta_n = \Delta(a_1, \dots, a_n)$, define the *adjacent cylinder set* Δ_n^+ to be $\Delta_n^+ = \Delta(a_1, \dots, a_{n-1}, a_n + 1)$. One can check (see Problem 12) that $\lambda(\Delta_n) \le 3\lambda(\Delta_n^+)$.

Next, suppose that I is an interval in $[0,1)$, and let $\Delta_n = \Delta(a_1, \dots, a_n)$ be the smallest cylinder set containing I. Then for almost all $x \in I$, there exists a j with $1 \le j \le 3$ such that either $\Delta_{n+j}(x) \subseteq I$, or $\Delta_{n+j}^+(x) \subseteq I$. This is because, depending on the parity of n, the cylinders at the $(n+1)^{\text{st}}$ level refine Δ_n either from right to left, or from left to right; see Figure 11.1.

Finally, we have all the ingredients needed to prove Lochs's Theorem.

Proof of Lochs's Theorem. Let's write S for the decimal map $S(x) = 10x - \lfloor 10x \rfloor$, D_n for its nth cylinder sets, and $D_n(x)$ for the nth cylinder set containing x.

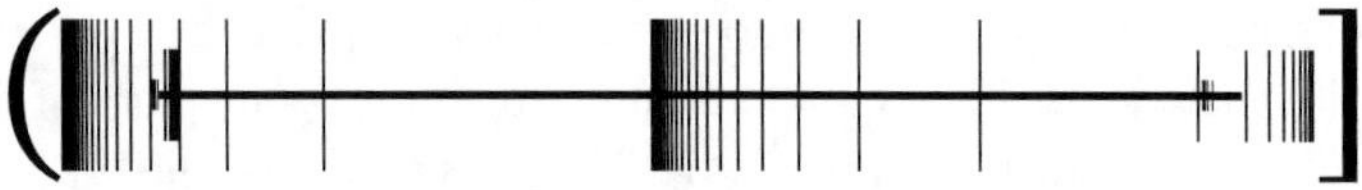

Figure 11.1. If Δ_n is the smallest cylinder set containing an interval I, then for almost all $x \in I$, either $\Delta_{n+j}(x) \subseteq I$, or $\Delta^+_{n+j}(x) \subseteq I$.

For $x \in [0, 1)$, find two cylinder sets $\Delta_m = \Delta_m(x)$ and Δ_{m+j} with

$$\Delta_{m+j} \subseteq D_n(x) \subseteq \Delta_m(x),$$

where Δ_{m+j} is either $\Delta_{m+j}(x)$ or $\Delta^+_{m+j}(x)$. Then we have

$$-\frac{1}{n}\log 3 + \frac{1}{n}\log \lambda(\Delta_{m+j}(x)) \leq \frac{1}{n}\log \lambda(D_n(x)) \leq \frac{1}{n}\log \lambda(\Delta_m(x)).$$

Now, $\log \lambda(D_n) = -n \log 10$, so for any $\varepsilon > 0$, for sufficiently large n we have

$$-\frac{1}{n}\log 3 - \frac{m(n,\alpha)}{n}\frac{\pi^2}{6\log 2} - \varepsilon \leq -\log 10 \leq -\frac{m(n,\alpha)}{n}\frac{\pi^2}{6\log 2} + \varepsilon.$$

Taking the limit gives

$$\lim_{n\to\infty} \frac{m(n,\alpha)}{n} = \frac{6\log 2 \cdot \log 10}{\pi^2}. \qquad \blacksquare$$

What this shows is that, for almost all numbers, a term of the continued fraction expansion provides slightly more information than a digit of the decimal expansion, because $\frac{6\log 2 \cdot \log 10}{\pi^2} < 1$. If we change from base 10 to base b, the corresponding ratio is $\frac{6\log 2 \cdot \log b}{\pi^2}$. If $b = 11$, then this ratio is slightly greater than 1, so a term in the continued fraction expansion gives slightly less information than a base-11 digit.

11.6 Problems

(1) Show that the largest possible entropy of a partition of X into n atoms is $\log n$.

(2) Let $(X, \mathcal{A}, \mu, T)$ be a measure-preserving system. Show that if $\mathcal{P}$ and $\mathcal{Q}$ are partitions of X, then $\mathcal{P} \vee \mathcal{Q}$ and $T^{-1}\mathcal{P}$ are also partitions of X.

(3) Prove Proposition 11.6.

(4) Show that if $\mathcal{P}$ and $\mathcal{Q}$ are independent, then $H(\mathcal{P} \mid \mathcal{Q}) = H(\mathcal{P})$.

(5) Prove Proposition 11.8.

(6) Prove Proposition 11.13.

(7) Show that if $\mathcal{P} \leq \mathcal{Q}$ and T is a transformation, then $h(\mathcal{P}, T) \leq h(\mathcal{Q}, T)$.

(8) Show that if $\mathcal{P}$ is a partition and T is a transformation, then for every positive integer n, we have

$$h(\mathcal{P}, T) = h\left(\bigvee_{i=0}^{n-1} T^{-i}\mathcal{P}, T\right).$$

(9) We saw that rational rotations on $\mathbb{R}/\mathbb{Z}$ have zero entropy. What about irrational rotations?

(10) Let T be the Lüroth series transformation

$$T(x) = \begin{cases} n(n+1)x - n & \frac{1}{n+1} \leq x < \frac{1}{n}, \\ 0 & x = 0 \end{cases}$$

on $\mathbb{R}/\mathbb{Z}$. Compute $h(T)$.

(11) Show that if $\mathcal{P}$ is a partition and T is a transformation, then $I_{T^{-1}\mathcal{P}} = I_{\mathcal{P}} \circ T$.

(12) With notation as in §11.5, show that $\lambda(\Delta_n) \leq 3\lambda(\Delta_n^+)$.

(13) Let $(\mathbb{R}/\mathbb{Z}, \mathcal{L}, \mu, T)$ be a measure-preserving system such that $\mu \ll \lambda$ and T is differentiable almost everywhere. Show that the entropy of T is given by

$$h(T) = \int_{\mathbb{R}/\mathbb{Z}} \log|T'|\, d\mu.$$

This is a special case of *Rokhlin's formula*. See [**VO16**, Theorem 9.7.3, pp. 296–299] for the full statement and a proof.

(14) Use Rokhlin's formula to give another proof that if T is the continued fraction transformation, then $h(T) = \frac{\pi^2}{6\log 2}$.

Appendix A

The Weak* Topology

A.1 Topological Spaces

Several proofs that we have seen, such as the Krylov–Bogolyubov Theorem, rely on facts about the weak* topology. In the proof of the Krylov–Bogolyubov Theorem, we need to find a T-invariant measure, and we do this by finding a convergent sequence of measures for each measurable set A. Then we somehow have to find a sequence that works simultaneously for all measurable sets A, and it is a fact about the weak* topology that guarantees that this is always possible. In this appendix, we will define the weak* topology and discuss the relevant facts about it needed to fill in the gaps in these proofs.

We begin by discussing topological spaces in general.

Definition A.1. Let X be a set. A collection $\mathcal{T}$ of subsets of X is said to be a *topology* if it satisfies the following properties:

- $\varnothing, X \in \mathcal{T}$.
- If A is any index set and $\{U_\alpha : \alpha \in A\}$ is a collection of sets in $\mathcal{T}$, then $\bigcup_{\alpha \in A} U_\alpha \in \mathcal{T}$.
- If n is a positive integer and $U_1, \ldots, U_n \in \mathcal{T}$, then $U_1 \cap U_2 \cap \cdots \cap U_n \in \mathcal{T}$.

We call the subsets in $\mathcal{T}$ the *open sets* of the topology, and we call the pair $(X, \mathcal{T})$ a *topological space*. When $\mathcal{T}$ is understood, then we simply call X a topological space.

One of the most familiar examples of a topology comes from a metric space. If (X, d) is a metric space, then it gives rise to a topology: a set $U \subseteq X$ is open if for all $x \in U$, there exists an $\varepsilon > 0$, depending on x, such that if $d(x, y) < \varepsilon$, then $y \in U$. A topology that arises in this way, from some metric on X, is said to be *metrizable*. However, not all topological spaces are metrizable. One way to see this is to note that all metrizable topological spaces are *Hausdorff*:

Definition A.2. A topological space $(X, \mathcal{T})$ is said to be *Hausdorff* if, whenever x and y are distinct points in X, there exist open sets $U, V \in \mathcal{T}$ with $x \in U$, $y \in V$, and $U \cap V = \varnothing$.

A simple example of a non-Hausdorff topological space is the two-point space $X = \{a, b\}$, where the open sets are $\varnothing$, X, and $\{a\}$. Then a and b do not have disjoint open neighborhoods.

It is possible to give a complete characterization of when a topological space is metrizable. The most useful result along these lines is the Nagata–Smirnov Metrization Theorem; see [**Mun00**, Theorem 40.3, pp. 250–252]. However, we will not need this result.

We have a notion of convergent sequences in a topological space.

Definition A.3. Let X be a topological space, and let $x_1, x_2, \ldots, x \in X$. We say that the sequence $x_1, x_2, \ldots$ *converges* to x, and we write $x = \lim_{n\to\infty} x_n$, if for every open set U of X with $x \in U$, there exists some N such that whenever $n \geq N$, then $x_n \in U$.

When X is a metric space, this agrees with the usual notion of convergence in a metric space. In general topological spaces, the limit of a convergent sequence need not be unique, but we do have uniqueness in Hausdorff spaces.

A.2 Continuity

Topological spaces are the natural setting for discussing continuity. In the context of functions on $\mathbb{R}$, or between metric spaces, the definition of continuity is typically given in terms of ε and δ. Remarkably, there is a much simpler-looking definition in the setting of topological spaces that coincides with the metric space definition when the spaces are metric spaces.

Definition A.4. Let X and Y be topological spaces. A function $f : X \to Y$ is said to be *continuous* if for every open set $V \subseteq Y$, $f^{-1}(V)$ is an open subset of X.

We can use functions to define topologies on sets. If we have a topological space X, a set Y, and a function $f : X \to Y$, we can put a topology on Y by declaring that f is continuous. Of course, this works if Y has very few open sets: if the only open sets on Y are $\emptyset$ and Y, then their preimages in X are $\emptyset$ and X, which are automatically open. The more open sets we add to Y, the harder it is for f to be continuous. In fact, there is a largest topology we can put on Y such that f is continuous.

On the other hand, if X is a set and Y is a topological space, and $f : X \to Y$ is a function, we can put a topology on X by declaring that f is continuous. Now the opposite is true: if every subset of X is open, then f is automatically continuous, and the fewer open sets X has, the harder it is for f to be continuous. Indeed, there is a smallest topology on X such that f is continuous.

Definition A.5. Let X be a set, let Y be a topological space, and let $\{f_\alpha\}_{\alpha \in I}$ be a collection of functions from X to Y. The *weak topology* on X is the smallest topology on X making each f_α continuous.

More concretely, for every open set $U \subseteq Y$ and every f_α, the preimage $f_\alpha^{-1}(U)$ is an open set in X, and then arbitrary unions and finite intersections of such sets are also open; these are exactly the open sets in X in the weak topology.

Generally, the term "weak topology" is used when X is a topological vector space over $\mathbb{R}$. Let X^* be the set of continuous linear functionals $T : X \to \mathbb{R}$. Then the weak topology on X is the smallest topology on X such that each element of X^* is continuous. When X is finite-dimensional, this is the same as the usual topology (sometimes called the *norm topology*) on X, given by treating X as $\mathbb{R}^n$ for some n.

Let us see why the weak topology is weaker than the norm topology in the infinite-dimensional setting. Let us consider an infinite-dimensional normed vector space V, with norm $\|\cdot\|$; for concreteness, let

$V = \ell^1(\mathbb{R})$ be the space of sequences $(x_0, x_1, \ldots)$ of real numbers such that $\sum_{n=0}^{\infty} |x_n| < \infty$, where the norm $\|(x_0, x_1, \ldots)\|$ is $\sum_{n=0}^{\infty} |x_n|$. If we have a nonzero continuous functional $T : V \to \mathbb{R}$, then its kernel has codimension 1, so we can write V as a direct sum $\ker T \oplus W$, where W is a 1-dimensional subspace of V. Thus the preimage under T of any nonempty open set must contain translates of $\ker T$, i.e. sets of the form $v + \ker T$ for some $v \in V$. If we intersect finitely many (say m) such preimages of open sets in $\mathbb{R}$ under various different linear functionals, then we find that this intersection contains translates of a codimension-m subspace of V. Since any weakly open subset of V must be a union of such finite intersections, every nonempty weakly open subset of V must contain translates of a subspace of V of finite codimension.

On the other hand, this is not the case for subsets of V that are open in the norm topology. For instance, if $V = \ell^1(\mathbb{R})$, then $\{(x_0, x_1, \ldots) : \sum |x_n| < 1\}$ is open in the norm topology, but not in the weak topology.

A.3 The Weak* Topology

Suppose X is a topological vector space, and let X^* be the space of continuous linear functionals $T : X \to \mathbb{R}$. We call X^* the *dual space* of X. The weak* topology is a topology on X^*. An element of X^* is a functional $T : X \to \mathbb{R}$, so given an element $x \in X$, $T(x) \in \mathbb{R}$. Initially, we think of X as being a set of vectors and X^* as being a set of functions on X, but we can also take the opposite perspective: each element $x \in X$ defines a functional $\phi_x : X^* \to \mathbb{R}$, defined by $\phi_x(T) = T(x)$.

Definition A.6. With notation as above, the *weak* topology* on X^* is the smallest topology on X^* such that ϕ_x is continuous, for all $x \in X$.

Note that each ϕ_x is a functional on X^*, hence an element of $(X^*)^* = X^{**}$. If X is finite-dimensional, then X^{**} is the set of all ϕ_x's, so X^{**} is naturally isomorphic to X. However, if X is infinite dimensional, then there may be other continuous linear functionals that do not arise as ϕ_x's, and X^{**} need not be isomorphic to X.

Since X^* is also a topological vector space, we can put a weak topology on X^*, which is the smallest topology making all the elements of

X^{**} continuous. Since the set of ϕ_x's is strictly contained in X^{**}, there are more functions that have to be continuous in the weak topology on X^* than in the weak* topology, so it is possible that there are sets in X^* that are open in the weak topology but not in the weak* topology. Indeed, this happens for some topological vector spaces X: every weakly open set in X^* is weak* open, but the converse need not hold.

Indeed, if X is an infinite-dimensional space, then one can give examples of sets in X^* that are weakly open but not weak* open. Let us describe the closed hyperplanes in X^* in both the weak and weak* topologies. We will start with the weak topology. Let $\psi \in X^{**}$ and $a \in \mathbb{R}$. Then the hyperplane

$$\{T \in X^* : \psi(T) = a\}$$

is closed in the weak topology; all closed hyperplanes in the weak topology are of this form. On the other hand, such a hyperplane is only closed in the weak* topology if ψ has the form ϕ_x for some $x \in X$.

To show that there exists a ψ not of the form ϕ_x for some $x \in x$, let $X = c_0(\mathbb{R})$, the space of sequences in $\mathbb{R}$ that converge to 0. The dual X^* turns out to be $\ell^1(\mathbb{R})$, which is the space of bounded sequences. The dual X^{**} of $\ell^1(\mathbb{R})$ is $\ell^\infty(\mathbb{R})$, the space of bounded sequences. If we have some sequence $a = (a_0, a_1, \ldots) \in \ell^\infty(\mathbb{R})$, then it behaves as a functional on $\ell^1(\mathbb{R})$ as follows: let $b = (b_0, b_1, \ldots) \in \ell^\infty(\mathbb{R})$. Then we define a acting as a functional on b by

$$\langle a, b \rangle = \sum_{n=0}^{\infty} a_n b_n,$$

which is a convergent series, hence it gives a real number as the output. (See [**Con90**, Examples 5.8 and 5.9, p. 76].) So, let's take a sequence in $\ell^\infty(\mathbb{R}) \backslash c_0(\mathbb{R})$, such as the constant 1 sequence $a = (1, 1, 1, \ldots)$. Then the hyperplane $\{b \in \ell^1(\mathbb{R}) : \langle a, b \rangle = 0\}$, which consists of those $b \in \ell^1(\mathbb{R})$ such that $\sum_{n=0}^{\infty} b_n = 0$, is closed in the weak topology but not the weak* topology.

A.4 The Space of Borel Probability Measures

Let X be a compact metric space, which comes with a Borel σ-algebra $\mathcal{B}$. We can consider the space of all finite real-valued measures (i.e. the measure can take on arbitrary real values, rather than just positive values; this can be considered as a difference of two nonnegative measures) on $(X, \mathcal{B})$, which we'll call $\mathcal{M}(X)$. We can also restrict to the probability measures; we'll call the space of Borel probability measures $\mathcal{M}_1(X)$; this time we require that the measures take on nonnegative values.

How does this fit in with our framework so far? We can start with the space V of continuous functions $f : X \to \mathbb{R}$. Then V forms a normed vector space, with norm given by $\|f\| = \sup_{x\in X} |f(x)|$. Since X is assumed to be compact, the supremum is finite, and is achieved by some $x \in X$. It turns out that we have a very neat description for V^*.

Theorem A.7 (Riesz–Markov–Kakutani Representation Theorem). *Let X be a compact Hausdorff space and ϕ a positive linear functional (i.e. $\phi(f) \geq 0$ for all $f \in V$) on V. Then there exists a unique Borel measure μ on X such that*

$$\phi(f) = \int_X f \, d\mu$$

for all $f \in V$.

In other words, $V^* = \mathcal{M}(X)$. See [**Rud87**, Theorem 2.14, pp. 40–47] for a proof.

Thus we are free to talk about the weak* topology on $\mathcal{M}(X)$, or $\mathcal{M}_1(X)$. The open sets on $\mathcal{M}_1(X)$ are defined to be the intersections of open sets in $\mathcal{M}(X)$ with $\mathcal{M}_1(X)$.

The next thing we need to discuss is a concrete interpretation of convergence of a sequence of probability measures in $\mathcal{M}_1(X)$.

Lemma A.8. *A sequence $\mu_1, \mu_2, \ldots$ in $\mathcal{M}_1(X)$ converges to $\mu \in \mathcal{M}_1(X)$ in the weak* topology if and only if for all measurable $f : X \to \mathbb{R}$, we have*

$$\lim_{n\to\infty} \int_X f \, d\mu_n = \int_X f \, d\mu.$$

See [**VO16**, Lemma 2.1.1, pp. 36–37] for a proof of a very similar theorem. In fact, it suffices that the limit holds whenever $f = \mathbb{1}_A$ for some Borel set A.

Our next goal is to show that $\mathcal{M}_1(X)$ is compact in the weak* topology. The first step in that direction is a key theorem from functional analysis.

Definition A.9. Let V be a normed vector space. The *operator norm* on the space V^* of continuous linear functionals of V is defined by

$$\|T\| = \sup_{\substack{v \in V \\ \|v\|=1}} \|Tv\|.$$

The unit ball in V^*, then, is the set of functionals $T \in V^*$ such that $\|T\| \le 1$.

Theorem A.10 (Banach–Alaoglu). *If V is a normed vector space, then the unit ball in V^* is compact in the weak* topology.*

See [**Lax02**, Theorem 3, pp. 120–121] for a proof.

We want to show that $\mathcal{M}_1(X)$ is compact in the weak* topology. This isn't immediately implied by the Banach–Alaoglu Theorem, because $\mathcal{M}_1(X)$ is not the unit ball in $\mathcal{M}(X)$. Indeed, it's contained in the unit *sphere* of $\mathcal{M}(X)$: the set of measures with operator norm 1; here the operator norm of a measure μ is $\sup_{A \in \mathcal{A}} |\mu(A)|$, which in the case of a positive measure is just $\mu(X)$. But $\mathcal{M}_1(X)$ is not the full unit sphere either, because we have the further restriction that for $\mu \in \mathcal{M}_1(X)$, we have $\mu(A) \ge 0$ for all $A \in \mathcal{A}$. So $\mathcal{M}_1(X)$ is in fact a very small subset of the unit ball (or sphere) in $\mathcal{M}(X)$.

It is worth noting that, while the unit *ball* is compact in the weak* topology, the unit *sphere* need not be. One way to see this is to note that in a compact metric space X, every sequence has a convergent subsequence. If we let $X = \ell^2(\mathbb{R})$ be the set of sequences $(a_0, a_1, \ldots)$ of real numbers such that $\sum_{n=0}^{\infty} a_n^2 < \infty$, then X is a normed vector space with norm

$$\|(a_0, a_1, \ldots)\| = \left(\sum_{n=0}^{\infty} a_n^2 \right)^{1/2},$$

and X is its own dual. The unit sphere in $\ell^2(\mathbb{R})$ is not compact in the weak* topology, because the sequence $e_0, e_1, e_2, \ldots$, where e_i is the sequence with a 1 in the ith position and 0's everywhere else, is a sequence in the unit sphere in $\ell^2(\mathbb{R})$ that has no subsequence that converges in the unit sphere. (The sequence converges to the 0 sequence in the weak* topology in $\ell^2(\mathbb{R})$, but the 0 sequence is not in the unit sphere.)

This may feel a bit unsettling in view of the fact that closed subsets of compact sets are compact. Additionally, the unit sphere is the preimage of 1 under the function $\|\cdot\| : \ell^2(\mathbb{R}) \to \mathbb{R}$ given by $\|(a_0, a_1, \ldots)\| = \left(\sum_{n=1}^{\infty} a_n^2\right)^{1/2}$, and the preimage of a closed set under a continuous function is closed. However, the operator norm function $\|\cdot\|$ is not continuous in the weak* topology, and that is where this "argument" breaks down.

But it is possible to modify this argument a bit to show that $\mathcal{M}_1(X)$ is compact in the weak* topology. First, we have to restrict to the subset of $\mathcal{M}(X)$ consisting of nonnegative measures. For each $A \in \mathcal{A}$, we require that $\mu(A) \geq 0$. The functional $\psi_A : \mathcal{M}(X) \to \mathbb{R}$ defined by $\psi_A(\mu) = \mu(A)$ is continuous, so the preimage of $[0, \infty)$, which consists of those measures μ for which $\mu(A) \geq 0$, is a closed subset of $\mathcal{M}(X)$. Since the intersection of closed sets is closed, and a closed subset of a compact set is compact, it follows that

$$C = \{\mu \in \mathcal{M}(X) : \mu(X) \leq 1 \text{ and } \mu(A) \geq 0 \text{ for all } A \in \mathcal{A}\}$$

is compact in the weak* topology.

This still isn't quite $\mathcal{M}_1(X)$; for instance, C contains the 0 measure. But there is one more condition we can require, which is that $\mu(X) = 1$. So, we can look at $\{\mu \in C : \mu(X) = 1\}$. Since $\{\mu \in \mathcal{M}(X) : \mu(X) = 1\}$ is closed, it follows that $\{\mu \in C : \mu(X) = 1\}$ is compact. But this last set is just $\mathcal{M}_1(X)$, so it follows that $\mathcal{M}_1(X)$ is weak*-compact.

Next, we need to know that $\mathcal{M}_1(X)$ with the weak* topology is metrizable. We show this by producing a metric on $\mathcal{M}_1(X)$ that induces the weak* topology. This metric is known as the *Lévy–Prokhorov metric*.

Definition A.11. Let (X, d) be a metric space, and let $\varepsilon > 0$. For $A \subseteq X$, let

$$A^\varepsilon = \{x \in X : d(x, a) < \varepsilon \text{ for some } a \in A\} = \bigcup_{a \in A} B_\varepsilon(a),$$

where $B_\varepsilon(a)$ is the ε-ball in X centered at a.

Definition A.12. Let (X, d) be a metric space with Borel σ-algebra $\mathcal{B}$. The *Lévy–Prokhorov distance* $\pi(\mu, \nu)$ be two Borel probability measures

$\mu, \nu \in \mathcal{M}_1(X)$ is defined to be

$$\pi(\mu, \nu) = \inf\{\varepsilon > 0 : \mu(A) \leq \nu(A^\varepsilon) + \varepsilon \\ \text{and } \nu(A) \leq \mu(A^\varepsilon) + \varepsilon \text{ for all } A \in \mathcal{B}\}.$$

Theorem A.13. *If X is a compact metric space, then the Lévy–Prokhorov metric on $\mathcal{M}_1(X)$ induces the weak*-topology.*

See [**VO16**, Proposition 2.1.4, pp. 39–40] for a proof.

Now, a metric space is compact if and only if every sequence has a convergent subsequence. Since we know that $\mathcal{M}_1(X)$ is a compact metric space, every sequence of Borel probability measures has a weak*-convergent subsequence. What it means for a sequence $\mu_1, \mu_2, \ldots$ of Borel probability measures to converge to some $\mu \in \mathcal{M}_1(X)$ is that for all Borel sets A, we have

$$\lim_{n\to\infty} \mu_n(A) = \mu(A).$$

This is just what we need to complete the proof of the Krylov–Bogolyubov Theorem, for instance.

Bibliography

[Bir31] George David Birkhoff, *Proof of the ergodic theorem.*, Proc. Natl. Acad. Sci. USA **17** (1931), 656–660 (English).

[BM80] Richard P. Brent and Edwin M. McMillan, *Some new algorithms for high-precision computation of Euler's constant*, Math. Comp. **34** (1980), no. 149, 305–312, DOI 10.2307/2006237. MR551307

[BM16] Vitaly Bergelson and Joel Moreira, *Van der Corput's difference theorem: some modern developments*, Indag. Math. (N.S.) **27** (2016), no. 2, 437–479, DOI 10.1016/j.indag.2015.10.014. MR3479166

[Bor09] Émile Borel, *Les probabilités denombrables et leurs applications arithmétiques.*, Rend. Circ. Mat. Palermo **27** (1909), 247–271 (French).

[Bre57] Leo Breiman, *The individual ergodic theorem of information theory*, Ann. Math. Statist. **28** (1957), 809–811, DOI 10.1214/aoms/1177706899. MR92710

[CG93] Lennart Carleson and Theodore W. Gamelin, *Complex dynamics*, Universitext: Tracts in Mathematics, Springer-Verlag, New York, 1993, DOI 10.1007/978-1-4612-4364-9. MR1230383

[Cha33] David Gawen Champernowne, *The construction of decimals normal in the scale of ten*, J. London Math. Soc. **8** (1933), no. 4, 254–260, DOI 10.1112/jlms/s1-8.4.254. MR1573965

[CK04] Marek Capiński and Ekkehard Kopp, *Measure, integral and probability*, 2nd ed., Springer Undergraduate Mathematics Series, Springer-Verlag London, Ltd., London, 2004, DOI 10.1007/978-1-4471-0645-6. MR2077699

[Coh06] Henry Cohn, *A short proof of the simple continued fraction expansion of e*, Amer. Math. Monthly **113** (2006), no. 1, 57–62, DOI 10.2307/27641837. MR2202921

[Con90] John B. Conway, *A course in functional analysis*, 2nd ed., Graduate Texts in Mathematics, vol. 96, Springer-Verlag, New York, 1990. MR1070713

[Daj14] Karma Dajani, *Introduction to ergodic theory and its applications to number theory*, `https://webspace.science.uu.nl/~kraai101/NotesWurzburg2014.pdf`, 2014.

[DK02] Karma Dajani and Cor Kraaikamp, *Ergodic theory of numbers*, Carus Mathematical Monographs, vol. 29, Mathematical Association of America, Washington, DC, 2002. MR1917322

[Erd81] P. Erdős, *On the combinatorial problems which I would most like to see solved*, Combinatorica **1** (1981), no. 1, 25–42, DOI 10.1007/BF02579174. MR602413

[EW11] Manfred Einsiedler and Thomas Ward, *Ergodic theory with a view towards number theory*, Graduate Texts in Mathematics, vol. 259, Springer-Verlag London, Ltd., London, 2011, DOI 10.1007/978-0-85729-021-2. MR2723325

[Fek23] Mihály Fekete, *On the distribution of roots of algebraic equations with integral coefficients*, Math. Z. **17** (1923), 228–249 (German).

[Fur77] Harry Furstenberg, *Ergodic behavior of diagonal measures and a theorem of Szemerédi on arithmetic progressions*, J. Analyse Math. **31** (1977), 204–256, DOI 10.1007/BF02813304. MR498471

[Fur81] Harry Furstenberg, *Poincaré recurrence and number theory*, Bull. Amer. Math. Soc. (N.S.) **5** (1981), no. 3, 211–234, DOI 10.1090/S0273-0979-1981-14932-6. MR628658

[GG55] Robert Ewing Greenwood and Andrew Mattei Gleason, *Combinatorial relations and chromatic graphs*, Canadian J. Math. **7** (1955), 1–7, DOI 10.4153/CJM-1955-001-4. MR67467

[Gow01] William Timothy Gowers, *A new proof of Szemerédi's theorem*, Geom. Funct. Anal. **11** (2001), no. 3, 465–588, DOI 10.1007/s00039-001-0332-9. MR1844079

[GT08] Ben Green and Terence Tao, *The primes contain arbitrarily long arithmetic progressions*, Ann. of Math. (2) **167** (2008), no. 2, 481–547, DOI 10.4007/annals.2008.167.481. MR2415379

[Kac47] Mark Kac, *On the notion of recurrence in discrete stochastic processes*, Bull. Amer. Math. Soc. **53** (1947), 1002–1010, DOI 10.1090/S0002-9904-1947-08927-8. MR22323

[KB37] Nicolay Kryloff and Nicolay Bogoliouboff, *La théorie générale de la mesure dans son application à l'étude des systèmes dynamiques de la mécanique non linéaire* (French), Ann. of Math. (2) **38** (1937), no. 1, 65–113, DOI 10.2307/1968511. MR1503326

[Khi64] Aleksandr Yakovlevich Khinchin, *Continued fractions*, University of Chicago Press, Chicago, Ill.-London, 1964. MR161833

[Khi98] Aleksandr Yakovlevich Khinchin, *Three pearls of number theory*, Dover Publications, Inc., Mineola, NY, 1998. Translated from the Russian by F. Bagemihl, H. Komm, and W. Seidel; Reprint of the 1952 translation. MR1616152

[Kri70] Wolfgang Krieger, *On entropy and generators of measure-preserving transformations*, Trans. Amer. Math. Soc. **149** (1970), 453–464, DOI 10.2307/1995407. MR259068

[Lax02] Peter D. Lax, *Functional analysis*, Pure Appl. Math., Wiley-Intersci. Ser. Texts Monogr. Tracts, Chichester: Wiley, 2002 (English).

[Lév29] Paul Levy, *Sur les lois de probabilité dont dependent les quotients complets et incomplets d'une fraction continue* (French), Bull. Soc. Math. France **57** (1929), 178–194. MR1504948

[Loc63] Gustav Lochs, *Die ersten* 968 *Kettenbruchnenner von* π (German), Monatsh. Math. **67** (1963), 311–316, DOI 10.1007/BF01299581. MR158507

[Loc64] Gustav Lochs, *Vergleich der Genauigkeit von Dezimalbruch und Kettenbruch* (German), Abh. Math. Sem. Univ. Hamburg **27** (1964), 142–144, DOI 10.1007/BF02993063. MR162753

[McM53] Brockway McMillan, *The basic theorems of information theory*, Ann. Math. Statistics **24** (1953), 196–219, DOI 10.1214/aoms/1177729028. MR55621

[Mul97] Irene Mulvey, *Recurrent ideas in number theory: the multiple Birkhoff recurrence theorem used to prove van der Waerden's theorem*, Math. Mag. **70** (1997), no. 5, 358–361. MR1573277

[Mun00] James R. Munkres, *Topology*, Prentice Hall, Inc., Upper Saddle River, NJ, 2000. Second edition of [MR0464128]. MR3728284

[Orn60] Donald S. Ornstein, *On invariant measures*, Bull. Amer. Math. Soc. **66** (1960), 297–300, DOI 10.1090/S0002-9904-1960-10478-8. MR146350

[Oxt52] John C. Oxtoby, *Ergodic sets*, Bull. Amer. Math. Soc. **58** (1952), 116–136, DOI 10.1090/S0002-9904-1952-09580-X. MR47262

[Pug15] Charles C. Pugh, *Real mathematical analysis*, 2nd ed., Undergraduate Texts in Mathematics, Springer, Cham, 2015, DOI 10.1007/978-3-319-17771-7. MR3380933

[Rot53] Klaus Friedrich Roth, *On certain sets of integers*, J. London Math. Soc. **28** (1953), 104–109, DOI 10.1112/jlms/s1-28.1.104. MR51853

[Rub23] Simon Rubinstein-Salzedo, *Transition to proofs*, World Scientific, 2023.

[Rud87] Walter Rudin, *Real and complex analysis*, 3rd ed., McGraw-Hill Book Co., New York, 1987. MR924157

[Sár78] András Sárközy, *On difference sets of sequences of integers. III*, Acta Math. Acad. Sci. Hungar. **31** (1978), no. 3-4, 355–386, DOI 10.1007/BF01901984. MR487031

[Sch16] Issai Schur, *Über die Kongruenz* $x^m + y^m \equiv z^m \pmod{p}$, Jahresber. Dtsch. Math.-Ver. **25** (1916), 114–117 (German).

[Sha48] Claude Elwood Shannon, *A mathematical theory of communication*, Bell System Tech. J. **27** (1948), 379–423, 623–656, DOI 10.1002/j.1538-7305.1948.tb01338.x. MR26286

[Sil07] Joseph H. Silverman, *The arithmetic of dynamical systems*, Graduate Texts in Mathematics, vol. 241, Springer, New York, 2007, DOI 10.1007/978-0-387-69904-2. MR2316407

[Sil08] Cesar E. Silva, *Invitation to ergodic theory*, Student Mathematical Library, vol. 42, American Mathematical Society, Providence, RI, 2008, DOI 10.1090/stml/042. MR2371216

[Soi09] Alexander Soifer, *The mathematical coloring book: Mathematics of coloring and the colorful life of its creators*, Springer, New York, 2009. With forewords by Branko Grünbaum, Peter D. Johnson, Jr., and Cecil Rousseau. MR2458293

[Sze75] Endre Szemerédi, *On sets of integers containing no k elements in arithmetic progression*, Acta Arith. **27** (1975), 199–245, DOI 10.4064/aa-27-1-199-245. MR369312

[vdC31] Johannes Gaultherus van der Corput, *Diophantische Ungleichungen. I. Zur Gleichverteilung Modulo Eins* (German), Acta Math. **56** (1931), no. 1, 373–456, DOI 10.1007/BF02545780. MR1555330

[VO16] Marcelo Viana and Krerley Oliveira, *Foundations of ergodic theory*, Cambridge Studies in Advanced Mathematics, vol. 151, Cambridge University Press, Cambridge, 2016, DOI 10.1017/CBO9781316422601. MR3558990

[Wil91] David Williams, *Probability with martingales*, Cambridge Mathematical Textbooks, Cambridge University Press, Cambridge, 1991, DOI 10.1017/CBO9780511813658. MR1155402

[Zak66] Elias Zakon, *Integration of non-measurable functions*, Canad. Math. Bull. **9** (1966), 307–330, DOI 10.4153/CMB-1966-040-5. MR200408

[Zak74] Elias Zakon, *Integration of non-measurable functions. II*, Canad. Math. Bull. **17** (1974), no. 4, 587–592, DOI 10.4153/CMB-1974-105-0. MR390157

Index